WIRE, CABLE, AND FIBER OPTICS FOR VIDEO AND AUDIO ENGINEERS

OTHER McGRAW-HILL BOOKS OF INTEREST

Bartlett   *Cable Communications*
Benson   *Audio Engineering Handbook*
Benson and Whitaker   *Television and Audio Handbook*
Coombs   *Printed Circuits Handbook, 4/e*
Croft and Summers   *American Electrician's Handbook*
Davidson   *TV Repair For Beginners, 5/e*
Fink and Beaty   *Standard Handbook for Electrical Engineers*
Fink and Christiansen   *Electronics Engineers' Handbook, 4/e*
Harper   *Handbook of Electronic Packaging and Interconnection, 2/e*
Inglis and Luther   *Video Engineering, 2/e*
Johnson   *Antenna Engineering Handbook, 3/e*
Lenk   *Lenk's Audio Handbook*
Lenk   *Lenk's Video Handbook, 2/e*
Mee and Daniel   *Magnetic Recording Handbook, 2/e*
Mee and Daniel   *Magnetic Recording Technology, 2/e*
Robin and Paulin   *Digital Television Fundamentals*
Rohde and Whitaker   *Communication Receivers, 2/e*
Sherman   *CD-ROM Handbook, 2/e*
Solari   *Digital Video and Audio Compression*
Whitaker   *Electronic Displays*
Whitaker   *DTV*
Williams and Taylor   *Electronic Filter Design Handbook, 2/e*
Yodes   *Home Audio*
Yodes   *Home Video*

Wire, Cable, and Fiber Optics for Video and Audio Engineers

Stephen H. Lampen

McGraw-Hill
New York · San Francisco · Washington, D.C.
Auckland · Bogotá · Caracas · Lisbon · London
Madrid · Mexico City · Milan · Montreal · New Delhi
San Juan · Singapore · Sydney · Tokyo · Toronto

Library of Congress Cataloging-in-Publication Data

Lampen, Stephen H.
   Wire, cable, and fiber optics for video and audio engineers /
  Stephen H. Lampen. — 3rd ed.
      p.   cm.
   Includes index.
   ISBN 0-07-037148-2 (hard edition). — ISBN 0-07-038134-8 (paper)
   1. Telecommunication wiring. 2. Telecommunication cables.
  I. Title.
TK5103.12.L36 1997
621.382'3—dc21

97-16148
CIP

McGraw-Hill

*A Division of The **McGraw·Hill** Companies*

1 2 3 4 5 6 7 8 9 0 DOC/DOC 9 0 2 1 0 9 8 7

ISBN 0-07-037148-2 (HC)   0-07-038134-8 (PBK)

*The sponsoring editor for this book was Steve Chapman, the editing supervisor was Sally
Glover, and the production supervisor was Pamela Pelton. It was set in Vendome ICG by
Jennifer L. Dougherty and Joanne Morbit of McGraw-Hill's Professional Book Group
composition unit, Hightstown, N.J.*

Printed and bound by R. R. Donnelley & Sons Company.

This book is printed on recycled, acid-free paper containing
a minimum of 50% recycled, de-inked fiber.

McGraw-Hill books are available at special quantity discounts to use as premi-
ums and sales promotions, or for use in corporate training programs. For more
information, please write to the Director of Special Sales, McGraw-Hill, 11 West
19th Street, New York, NY 10011. Or contact your local bookstore.

CONTENTS

Contents

Contents

Contents

ACKNOWLEDGMENTS

This book could not have been written without the help and encouragement of a number of people.

First and foremost is the tireless staff at Belden Wire & Cable, especially Doug Brenneke, Kip Coates, Greg Deitz, Wynn Roth, Will Stacy, Butch Thompson, Sheyla Watson, and especially Marty Van Der Burgt.

Second, many knowledgeable end users are represented herein, including Bill Ruck, Chief Engineer KFOG/KNBR, San Francisco, and Tim Pozar, Chief Engineer KKSF/KDFC, San Francisco.

Third, I am indebted to professional engineers and consultants including Pete Blackford, Richmond, Ind.; Mike Newman, CSI, San Francisco; and Dane Ericksen and Bill Hammett of Hammett & Edison, San Francisco.

And, last, thanks to the willing victims who read this manuscript and made honest comments to help make it a readable tome, including Ambrose Lampen, Michael Lampen, and my wife, Debra.

TRADEMARKS

Throughout this book, various trademarks are mentioned, which include the following:

- Teflon, Kevlar, Mylar, and Hypalon are trademarks of DuPont Corporation.
- Heliax is a trademark of Andrew Corporation.
- Solef is a trademark of Solvay Polymers, Inc.
- Halar is a trademark of Ausimont Corp.
- Kynar is a trademark of Pennwalt Corporation.
- Silicone is a trademark of General Electric.
- Ethernet is a trademark of Xerox Corporation.
- QCP is a trademark of ADC Telecommunications, Inc.
- Bantam, TT, and tini telephone are trademarks of Switchcraft, Inc.
- DirecTV is a trademark of DirecTV, Inc.
- VHS is a trademark of Panasonic, Inc.
- Flamarrest, Zetabon, DataTwist 350, and Duofoil are trademarks of Belden Wire & Cable.

Other manufacturers are also mentioned, including AMP, Neutrik, ITT Cannon, TRW/Cinch, and Molex.

1

The Basics

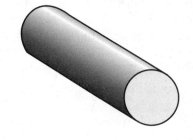

Let's say you have a battery and a light bulb (Fig. 1-1). A battery uses chemicals to produce electricity. A light bulb turns electricity into heat and, most important, light. But how to connect these two together?

If you can conduct the electricity from the battery to the light bulb, you will have light. Since there is a "+" and a "—" terminal on the battery, and also two connections on the light bulb, if you attach the "+" to one terminal and the "—" to the other, the electricity made by the battery will flow through the filament in the bulb and it will light up (see "Notes and Comments" appendix, item 1).

If you use two pieces of straw to connect them, nothing will happen. If you try two pieces of string, even though the connections were perfect, nothing will happen. These do not conduct electricity. You need a *conductor*. If you make two thin wires out of metal, which is a conductor of electricity, and attach them to the battery and light bulb, they look like Fig. 1-2.

And the light bulb lights up! Turning it on and off is a different matter. You have to unplug the light bulb or the battery, or cut the wire. You have to add a *switch* (Fig. 1-3).

When the switch is open, no electricity will flow. When the switch is closed, it completes the circuit and the light bulb lights. Since usually you want to place the light bulb on the ceiling to light the room, it would be easier to run the wires together, as in Fig. 1-4.

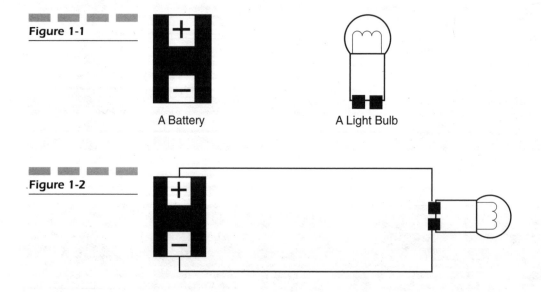

Figure 1-1

A Battery   A Light Bulb

Figure 1-2

Figure 1-3

Figure 1-4

Figure 1-5

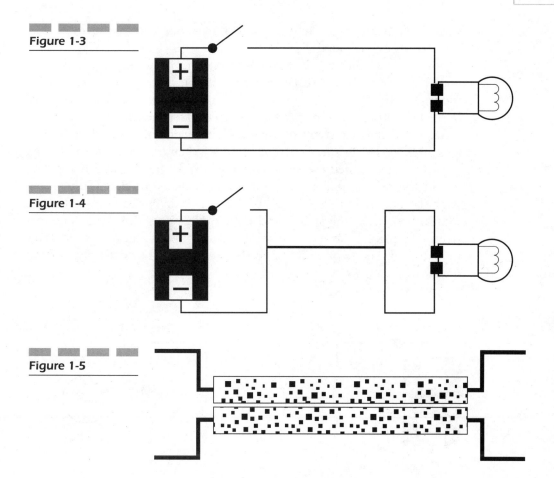

The light will not light, however, even with the switch closed. In fact, the wires will begin to get very, very hot. So you will turn off the circuit. You'll notice that the wires are touching each other. If you use bare wires, you must not let them touch; otherwise the electricity, which always takes the easiest path, will flow where they touch instead of through the light bulb (Fig. 1-6). In other words, the circuit is shorter than the one you tried to create. It's a *short circuit!* You must wrap each wire in a material that does not conduct electricity: an *insulator.*

These diagrams generally show bare wires, but they probably would be insulated except where the connections were made (Fig. 1-5).

You could make the wire *solid,* that is, one thick piece of metal. To increase flexibility, you could make it *stranded,* composed of many smaller wires (Fig. 1-7). The diagrams show solid wires, but they just as easily could be stranded.

You could put a light bulb in the next room but keep the switch in your own room. If you turned the switch on and off, someone in the room with the light bulb would see it flashing and know that you were in the other room. You had *communicated* to the other person.

We could make a graph of what the light turning on and off might look like (Fig. 1-8). First off, then on, then off, then on, then off, and so on...

Now if you turned the switch on and off like Fig. 1-9, and if the person in the other room could read Morse code, he or she would know your flashing meant "S.O.S." or "HELP!"

In other words, the switch and light could be used to send messages from one place to another. A simple series of on-and-off could send information. The light could be in the next room, the next house, or the next town.

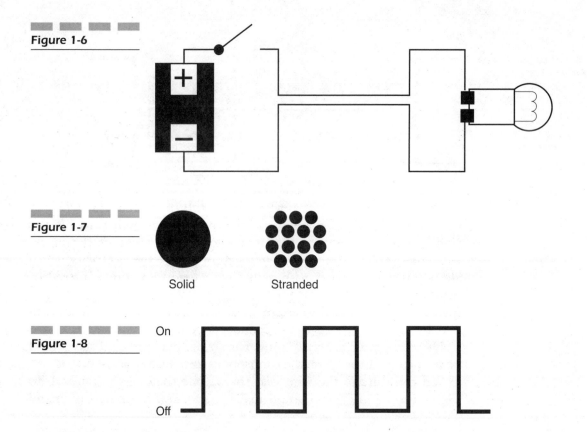

Figure 1-6

Figure 1-7

Solid   Stranded

Figure 1-8

On

Off

Figure 1-9

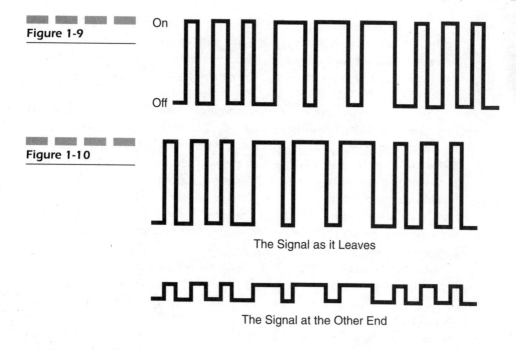

On

Off

Figure 1-10

The Signal as it Leaves

The Signal at the Other End

Figure 1-11

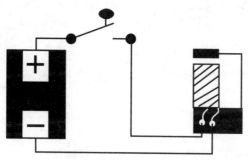

All you need to get to the next town is enough electricity to overcome the *resistance* in the wire, which wants to turn the electricity into heat. Resistance can turn a good signal into an unusable signal (Fig. 1-10).

You could use a wire with less resistance (copper instead of steel, for example), but that would be more expensive. You could use thicker wire of the same type (more conductive area would mean less resistance), but that also would be expensive. You could use more electricity (higher voltage), but that would be dangerous. Or you could leave everything as it is and invent a more sensitive receiver. Historically, the sensitive receiver won out (Fig 1-11).

The sensitive receiving device was a coil of wire with a piece of iron on top. When electricity runs down a wire, it creates or *induces* a small magnetic field around the wire; this is called *electromagnetism*. Electromagnetism is like regular magnetism, except it only appears when electricity is flowing.

Wires are said to have *inductance*, the ability to store electrical energy in a magnetic field around the wire (see "Notes and Comments," item 1). If the wire is coiled, magnetism is concentrated. Coils of wire are often called *inductors* because of this very property. When the switch in Fig. 1-11 is closed, the receiving coil becomes a magnet and attracts the piece of iron above it, making a click. This uses much, much less electricity than flashing a light bulb. Using the same battery and the same wire, a message can be sent farther with this arrangement than with a light bulb ("Notes and Comments," item 14).

Someone might tell you they could reduce the cost of the wire by half! The ground under our feet is a good conductor. Just put a heavy copper rod in the ground at each end of the circuit and run one long wire (Fig. 1-12), saving almost half the cost! (Also see "Notes and Comments," item 10.) Of course, that single wire could not touch the ground, because it also is a conductor. If it did touch, there would be a short circuit. You must put the single wire up on wooden poles: telegraph poles (Fig. 1-13).

Obviously you cannot drag your equipment and poles around. You needed some way to draw your inventions, to show the scheme of these

Figure 1-12

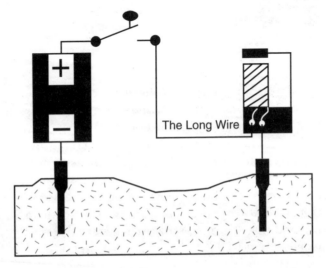

The Long Wire

Figure 1-13

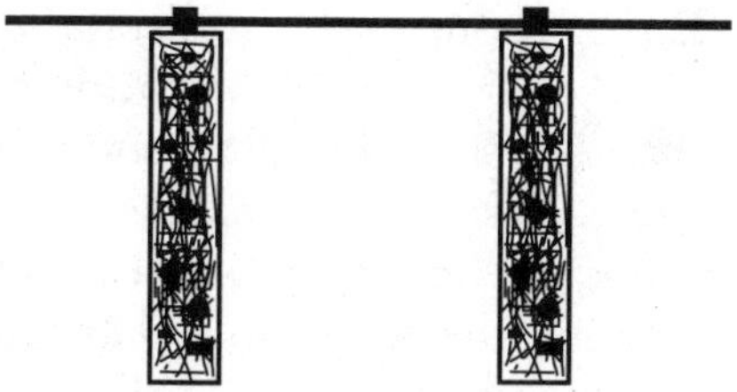

Figure 1-14

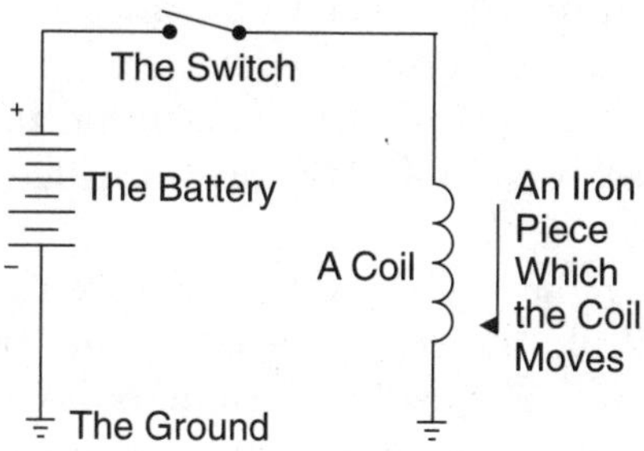

The Switch

+

The Battery

A Coil

An Iron
Piece
Which
the Coil
Moves

−

The Ground

Figure 1-15

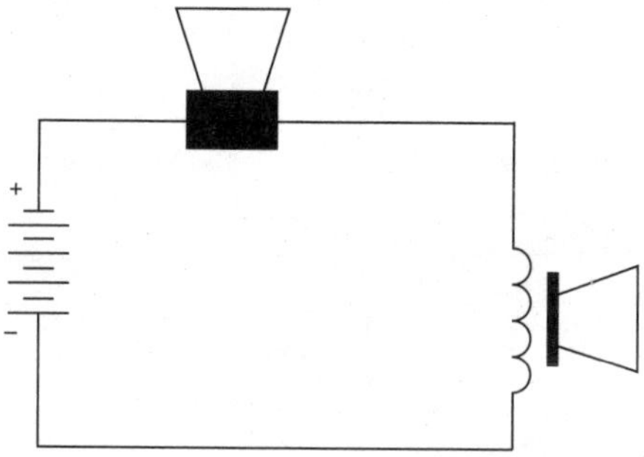

circuits—a *schematic.* Figure 1-14 is the schematic diagram of your tele-graph circuit.

You could replace the switch above with a cardboard funnel attached to a small container of carbon. You also might know from experimentation that vibrating the carbon changes the electrical flow through it. At the other end, you could attach a thin piece of iron to another funnel. When you yell into the microphone piece, your faint voice could be heard distinctly in the receiver (Fig. 1-15). You have invented the telephone! (See "Notes and Comments," item 2.)

Now you might think you could wire this telephone the same way you ran the telegraph (Fig. 1-16), but after running the wire only a few feet—not even to the next house—the noise on the receiver would be so great that you couldn't hear the person on the other end.

Something has changed! Strangely, the noise is worse during the day and better at night. Finally you would realize where the noise was coming from: the sun. You would notice another odd thing: When you hooked up the telephone the way you did before, with two wires, it worked much better, with less noise. When the two wires were run very close together, it worked very well, with very little noise. By accident, you have also invented something that would help reduce unwanted noise being heard in the receiver: a *balanced line.*

Here's how it works. With the microphone in one room and the receiver in the other, you would run two insulated wires from room to room as in Fig. 1-17.

If you had "magnetic vision," you could see the flow of electricity down the wires. While there is a constant flow from the battery, it is being controlled or *modulated* by the microphone as someone speaks into it. The way the electricity flows, even when nobody speaks into the

Figure 1-16

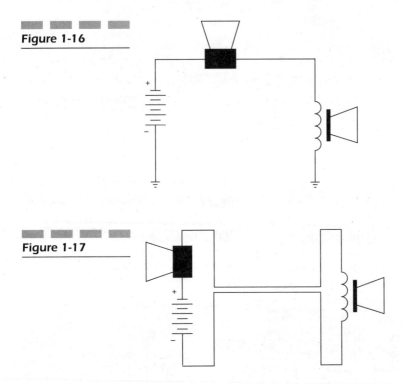

Figure 1-17

Figure 1-18

Figure 1-19

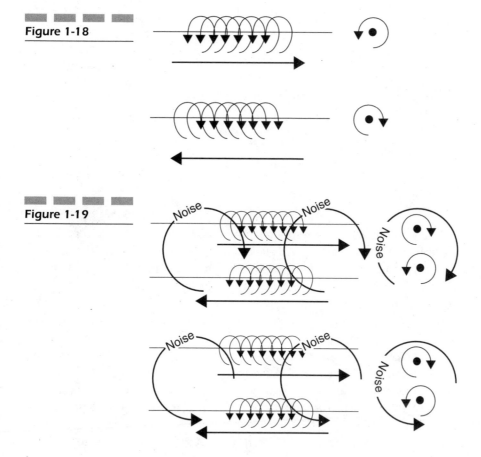

microphone, is in one direction on one wire and the other direction on the other wire (Fig 1-18).

Electrical flow produces a magnetic field around the wire. The direction of that field around that wire is determined by the direction of the flow. Electromagnetic noise that comes from outside our circuit travels only in one direction, either clockwise or counterclockwise (Fig. 1-19).

Any electromagnetic noise will induce a current to flow down the wires. The flow direction of the noise current is the same on both wires, but the receiver only reproduces signals that are in the opposite direction on the wires. Because the electromagnetic noise has the same direction on both wires, the noise meets itself in the receiving coil and cancels itself out! Only the original signal (the voice) is heard at the other end (Fig. 1-20).

Figure 1-20

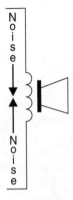

The only thing to remember is to keep the wires as close together as possible. If they are apart, the electromagnetic interference hits them at a slightly different time and angle and you hear that difference as noise. You can keep two wires next to each other all the time by twisting them together. Of course, these wires have to be insulated in some way to prevent the bare wires from touching one other, but that's easy to do. The *twisted pair* is born (Fig 1-21).

Surprisingly, the noise level is even lower! Because the wires not only are next to each other but also revolving around one another, their electromagnetic fields are also rotating. It becomes even more difficult for a given source of electromagnetic interference to affect the signal going down the wire.

In a balanced line, the vast majority of its noise immunity comes from the fact that it is a twisted pair. That's lucky, because interference comes from everywhere: electric motors, car engines, radio broadcasters and, of course, the sun. Interference signals fall into two classes. *Electromagnetic interference* (EMI) comes from sources like power lines and electric motors. *Radio frequency interference* (RFI) comes from many of these same sources, plus lightning, spark plugs, radio and television transmitting equipment, and computers.

There is one other interesting source of interference: other wires. As telephone service grows, more and more pairs are run side by side (Fig. 1-22).

If you ever hear someone else's conversation faintly in the background while you're on the phone, that's *crosstalk*. Twisted pairs are very good at rejecting crosstalk. That distant-sounding conversation probably is caused by a phone installer untwisting your wires, and someone else's wires, so he could terminate them. In untwisting them, he destroyed much of their ability to reject one another's signal. Kept twisted, balanced-line twisted pairs are extremely effective and are still the mainstay of the phone company. It was also found that twisting the pairs at different twist ratios (twists per inch) and twisting them right and left (clockwise and counterclockwise) reduced the crosstalk even further.

Though this might satisfy the phone company, there are other users who want even better performance, namely broadcasters and audio (sound) engineers. While the signal on a carbon microphone in a telephone is clear and strong, it is not of very good quality. It does not reproduce the whole audible spectrum, generally considered to be 20—20,000 cycles per second (or 20 Hz to 20 kHz). Audio people want to use something better: a *dynamic microphone*.

Such a microphone moves a coil of wire that's wound around a magnet to put the signal on the circuit ("Notes and Comments," item 2). Its conversion of sound to electricity is much more accurate than our previous design. This arrangement eliminates the battery in the microphone section; the microphone by its very motion generates the signal. The only problem is the level of the signal, sometimes as low as one-millionth that of the carbon microphone and battery. The signal on the line is very, very weak and very easily interfered with. It also means that the signal will not be able to drive a speaker by itself. The signal level will have to be raised or amplified in order to be used.

You can protect the smallest signal from interference by shielding the twisted pair. Around the twisted pair you put a braid of wire. At each

Figure 1-21

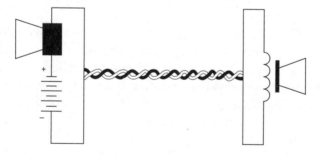

Figure 1-22

Figure 1-23

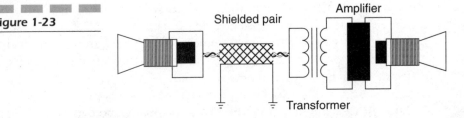

end you attach that braid to ground. Any interference that hits this shield is sent to ground. You also will find that the twisting of the pair reduces electromagnetic noise and the shield reduces electrostatic noise. You can attach the shield by unbraiding the ends and twisting those strands together to make a wire.

You also can shield the twisted pair with foil. Foil shielding is not as effective for EMI, but extremely effective on RFI. Foil covers the twisted pair 100 percent (which no braid can) and is much cheaper to produce. The mainstay of the broadcast industry is foil-shielded, stranded, twisted-pair wires. Since you cannot connect directly to the foil, a third uninsulated wire is added. This wire touches the foil and makes the necessary ground connection to the shield. It is called a *drain wire* because it allows you to drain off the noise and interference that hits the shield.

If you play with coils of wire, you will find that if there is a signal in one coil, it can magnetically pass that signal to a coil near it. The closer the other coil is, the stronger the transferred signal will be. This is called a *transformer*. Winding the wire around iron plates makes the coils better at passing the magnetic energy from one to the other. The two black lines in the transformer schematic symbol (Fig. 1-23) indicate that iron plates are inside.

When feeding balanced lines, a transformer allows the signals to transfer while still maintaining the noise-cancellation features of the balanced line. You can design circuits, called *active balanced lines*, that maintain balanced lines without the use of a transformer, but the transformer is still the easiest and most common solution.

In addition to dynamic microphones, you also can make dynamic speakers with magnets and coils, and get better, more accurate audio reproduction. Virtually all speakers today are made in this fashion.

Simply by playing with wire, you have created a number of items that are keys to all of today's marvels. So what is wire? And what is cable? What is it made of and how does it work?

2

Cable Parameters

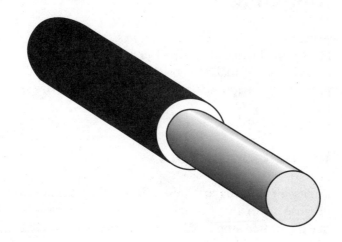

The Conductor

In order to conduct electricity, you need a conductor (Fig. 2-1). Metals are the least expensive, most available, and most easily workable materials for conductors. Wire is made with various conductive metals. Some metals are better conductors than others (Table 2.1).

The lower the number, the better the conductor. Some metals are easier and better to work with than others, as shown in Table 2.2.

Annealing

Annealing is a process that restores the flexibility of a wire after it has been pulled through a die, or *drawn*, from a larger gage to a smaller gage. Drawing wire makes it brittle (a property called *work hardening*) and prone to breakage. Annealing ovens heat the wire and restore its molecular flexibility.

Exotic Conductors

Purity of the material, especially silver and copper, can be an issue because it can affect conductivity. Some purists swear by OFC, or *oxygen-free copper* (see Appendix A, "Notes and Comments," item 5). The jury is still out on OFC and other exotic conductors. Some swear they can hear the difference, while others say it's hype. There is no laboratory evidence that, other than the basic elements and parameters outlined in this book, there is anything that truly makes a difference. That is not to say, however, that some future laboratory measurement or procedure won't show objective evidence of the superiority of OFC, or any similar material. It just hasn't happened yet.

Figure 2-1

TABLE 2.1

Circular mil-ohms per foot at 20°C	
Silver	9.9
Copper	10.4
Gold	14.7
Aluminum	17
Nickel	47
Steel	74

TABLE 2.2

Metal	Cost	Conductivity	Strength	Flexibility	Annealing
Silver	High	Excellent	Poor	Poor	Poor
Copper	Medium	Very Good	Good	Good	Good
Gold	Very High	Good	Poor	Excellent	Not needed
Aluminum	Fair	Low	Good	Good	Good
Nickel	Medium	Poor	Good	Poor	Poor
Steel	Very Low	Very Poor	Excellent	Excellent	Not needed

Wire Gage

We can make large-gage wires or small-gage wires. The standard sizes are specified as *American Wire Gage*, or *AWG*. A few different wire gages are shown in Table 2.3.

Solid and Stranded Wire

We can make one large conductor (solid wire), or we can make a conductor by wrapping together a number of smaller conductors (stranded wire). Table 2.4 shows two different stranded gages and several of their common strand constituents.

You can always use small strands to make larger-gage wires. As you increase the number of strands, the flexibility or limpness and the flex life increase, but so does the cost. *Flex life* is how long a wire will flex until it breaks.

Galvanic Corrosion and Oxidation

Metals have two problems, oxidation and corrosion. *Oxidation* (Fig. 2-2) is the combining of the metal with oxygen and other chemicals, such as sulfur, to create "rust." *Corrosion,* or *galvanic corrosion,* is the result of the electrical potential that can occur between dissimilar metals, sometimes eating away one metal or depositing the other on its surface. (This is the same process, by the way, that's put to positive use in a car battery.) Because oxidation and corrosion can affect the dimensions of a piece of metal, they can have a major effect on connector fit, current-carrying capacity, and conductivity—and therefore reliability. Table 2.5 is a list

TABLE 2.3

AWG	Diameter (in)	Compares to...
40	0.0031	Smaller than a human hair
30	0.0100	Sewing thread
20	0.0320	Diameter of a pin
10	0.1020	Knitting needle
1	0.3900	Pencil
1/0	0.4100	Finger (called "1-aught")
3/0	0.4640	Marking pen (called "3-aught")
4/0	0.6080	Towel rack (called "4-aught")

TABLE 2.4

Stranded AWG	can be made by	Strands	of	Gage
20	=	7	X	28
20	=	19	X	32
14	=	7	X	22
14	=	19	X	27
14	=	42	X	30

Figure 2-2

TABLE 2.5

Metal	Oxidation	Oxide Conductivity
Silver	Bad	Excellent, same conductivity
Copper	Poor	Poor, a semiconductor
Gold	Excellent	Excellent, no oxide
Aluminum	Bad	Bad
Nickel	Good	Good
Tin	Good	Good
Steel	Bad	Bad

Figure 2-3

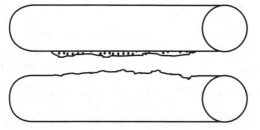

of different metals, how badly they oxidize, and the conductivity of the oxide.

Corrosion can be predicted by the electrical potential between dissimilar metals. All you need are two dissimilar metals and an acid between them. Even rainwater, fog, or salt spray is enough to start a corrosive reaction. Note that one conductor is added to while the other is taken from (Fig. 2-3). Both conductors will have compromised performance.

Table 2.6 shows the reaction potential of various metals compared to hydrogen as a 0-volt reference.

If you bond a less reactive metal to a more reactive one, the potential to corrode can be reduced. For instance, copper is often dipped in a bath of molten tin. The tin coating reduces the potential to corrode by almost two-thirds. Tin is not a good conductor, but is much less

TABLE 2.6

Metal	Reaction in Volts
Gold	+1.498
Silver	+0.799
Copper	+0.337
Hydrogen	0 (reference)
Tin	-0.136
Nickel	-0.250
Aluminum	-1.662

prone to oxidization than copper, and it's only a thin layer coated on the copper cable.

Whether a cable actually corrodes depends upon the quality of the jacket materials around the cable and the weathering, such as water- and airborne chemicals it may encounter where it is installed. Corrosion is rarely an issue for cable installed inside buildings.

In manufacturing, jacket materials are extruded (melted and squeezed) over the wires. Where this is not practical, or where wires will be stripped of their jackets and exposed when in use, we can coat wires with more resistant materials, such as tin over copper.

Where the environment is especially harsh, such as with immersible, direct-burial, or deep-sea cables, multiple jackets can be provided, sometimes filled with "gel" to prevent water or air leakage. They can even be constructed with aluminum or steel armor over the cable to help prevent leakage.

Corrosion and Connectors

Connectors are most often silver- or nickel-plated. If nickel is used, the potential between copper and nickel or silver and nickel can be quite substantial (over 1 volt). This can lead to corrosion and connector failure, especially outdoors. In those cases, care should be taken to use copper wire (or silver-clad copper) with a silver connector, thus reducing the galvanic potential to about half a volt. Better still, tinned wire in a nickel connector gives a reaction only slightly more than 0.1 volt.

If the wire is installed in a rack inside a building with air conditioning, the potential for corrosion or oxidization can be reduced greatly, probably beyond the lifetime of the equipment. This is also one factor that should be taken into consideration when reusing wire from an old installation.

Frequency and Performance

Many cable specifications vary with the frequency traveling down the wire. We can use any of various devices, such as an oscilloscope, to see these frequencies. Most signals you see will be alternating back and forth on the circuit, going from plus to minus. These are called *alternating current* or *ac.*

Alternating current travels in cycles. If it takes 1 second to go back and forth, from 0 to plus to minus and back to 0, then it is traveling at 1 cycle per second, also called 1 hertz (abbreviated Hz). How often these cycles occur is called the *frequency,* three examples of which are shown in Fig. 2-4. Oscilloscopes have the ability to let us see much higher frequencies, into the thousands or millions of hertz (called *kilohertz* and *megahertz,* respectively, and abbreviated kHz and MHz). Table 2.7 shows the frequencies of several common alternating phenomena.

If we adjusted the oscilloscope to show one-millionth of what was coming in, the same traces would actually be showing us 2, 10, and 20 MHz. In this way we can see and control frequencies much higher than we could normally sense.

TABLE 2.7

Use	Frequency (in hertz)
Lowest audible frequency	20
Power out of a wall socket	60
A on a piano (A_4)	440
Top key on a standard piano (C_8)	4186
Highest audible frequency	20,000 (20 kHz)
Middle of the AM radio band	1,100,000 (1100 kHz or 1.1 MHz)
Beginning of TV channels	54,000,000 (54 MHz)
Middle of the FM radio band	98,000,000 (98 MHz)

Figure 2-4

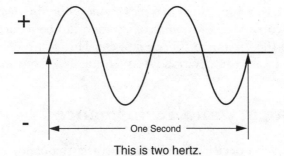

This is two hertz.

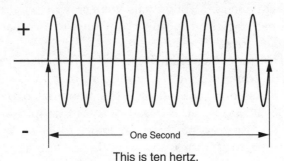

This is ten hertz.

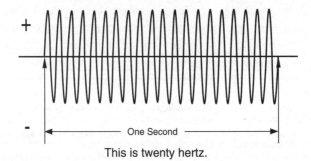

This is twenty hertz.

Wavelength

The traces in the illustration look like waves on a pond, and you often will read about the *wavelength* at a certain frequency. That is the distance between crests or between valleys when a wave of a given frequency rises, falls, and rises again. In other words, it's the physical distance covered by one cycle of a wave traveling in a particular direction.

It should be noted that there is an *electrical wavelength* (for electromagnetic waves) and an *acoustic wavelength* (for sound waves traveling through the air). The formula for acoustic wavelength is very simple:

$$\lambda = \frac{1127}{F}$$

where λ is the wavelength in feet, F is the frequency in hertz, and 1127 is the speed of sound in air at sea level, in feet per second. For example, the 66th key on a piano (D_6), which is 1175 Hz, has a wavelength slightly under a foot; the 25th key (A_2), with a frequency of 110 Hz, has a wavelength just over 10 ft.

We are interested in signals passing down wires, however, and therefore are concerned with the electromagnetic wavelength, the formula for which is equally simple:

$$\lambda = \frac{300,000,000}{F}$$

where λ is the wavelength in meters, F is the frequency in hertz, and 300,000,000 is the speed of light in meters per second.

Thus a signal of 300 MHz passing down a wire will have a wavelength of 1 m (3 ft 3 in), and a signal of 10 MHz will have a wavelength of 30 m (98 ft).

Skin Effect

Direct current flows down the entire conductor. But as ac signals pass down wire, and as their frequencies increase, the signals tend to travel down the surface of the wire. This is called *skin effect*.

Figure 2-5 shows the skin depth in mils on a tin, copper, or silver conductor. A *mil* is 1/1000th of an inch.

Because of skin effect, if a wire is designed for use only at higher frequencies, we need only coat the outside to be conductive. The inner wire can be almost anything, since there will be no signal passing down it. For high-frequency cable you might see a steel or aluminum center with a copper or silver coating. This is a clear indication that the wire is intended for high-frequency applications above 50 MHz. Some of the combinations that exist are:

- Silver-clad copper
- Copper-clad steel
- Silver-coated copper-clad steel

Because silver makes brittle wire but is the best conductor, it lends itself well to coating. Care must be taken, however, to make the assembly corrosion- and oxidation-resistant, as silver readily corrodes and oxidizes. At high frequencies, steel can be used in the center of the wire because the signal passes down the skin. Steel can add great strength to a wire, and it is cheap.

If wire is to be used for low frequencies only, such as audio, then it must be all copper because the entire conductor is being used. Note in Fig. 2-6 that the skin depth at 0.1 MHz (100 kHz) is 10 mils. At 20 kHz the skin depth is about 30 mils. An average audio cable has 22 AWG conductors. Solid 22 AWG is 25 mils in diameter, and stranded is about 30 mils. Therefore the skin depth at the highest audio frequencies is the entire cable. Still, some audio engineers claim they can hear the difference between copper and silver-clad copper at audio frequencies.

Figure 2-5

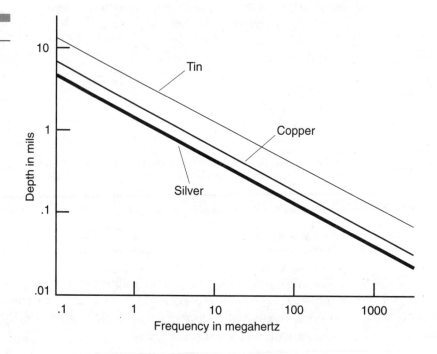

For wire to work well at both low and high frequencies, it must be all copper at least, or silver-coated copper at best, because you need the core for low frequencies and the skin for high frequencies.

There is one interesting effect that can occur when using tinned copper wire at higher frequencies. Since tin oxidizes, thus protecting the copper beneath, and since both the tin and its oxide are bad conductors, there is often little skin effect simply because of the high resistance of the tin.

In that case, most of the conductivity is still in the copper core. Thus a tinned copper wire cannot be dismissed for high-frequency use without acknowledging this effect. The suitability of a tinned wire in a nickel-plated connector may make such a choice suitable, especially in outdoor applications.

Insulators and Dielectrics

Wires can be left bare, but this promotes oxidation and corrosion, allows multiple wires to touch or short circuit, and makes them hazardous if significant voltage potentials exist between them. Therefore we coat the wires with various nonconductive compounds (Fig. 2-6). This layer over the wire is called *insulation*. If the electronic performance of the wire is affected by the nonconducting coating, it is described as a *dielectric*.

Any plastic can be used to coat a wire, but some perform better than others. Table 2.8 shows some of the common plastics used for communications-grade cables. (PVC is polyvinyl chloride.) *Colorability* describes how readily this plastic can be made in colored versions for color coding and identification. A is excellent, F is poor.

Extrusion

Nearly all wires are coated with plastic through a process known as *extrusion*. Plastic in the form of beads or pellets is melted under heat and squeezed into a small space that has a bare wire being pulled through it. When the wire exits the space, it is coated with the plastic. The wire is immediately immersed in a water bath to cool and harden the plastic. This is a continuous process, with the wire in motion the entire time.

Figure 2-6

TABLE 2.8

Plastic	Cost	Colorability	Flexibility	Flammability	Performance
PVC	A	A	A	C-A	D
Polyethylene	B	B	B	D	A
Teflon™	D	F	D	A	A

Thermoplastic and Thermoset

Materials used to coat wires come under two categories. First is *thermoplastic*. This term describes a material that can be heated, manipulated, and cooled, then heated and manipulated and cooled again. For instance, PVC is a thermoplastic. It may be heated to have color dyes added or special flame-resistant ingredients added. It then can be cooled and reheated later for extrusion. Some thermoplastic materials, such as Teflon, require high temperatures and special techniques to extrude. In that case, a manufacturer may resist using Teflon for a multistep project unless it is absolutely necessary.

Thermoset materials, such as rubber or EPDM (which is artificial rubber), can be heated and used only once. If colors are desired, for instance, the plastic must be manufactured initially with that color. Once heated, it changes its state and cannot be changed back again. Table 2.9 gives examples of each class of insulating material.

Insulation and Performance

At low frequencies (below 1000 Hz) and moderate voltages (below 600 volts), insulators have little or no electronic effect. It's just a question of how rugged, how flame-resistant, or how cheap an insulator you want.

This is why you see many materials (such as rubber, Neoprene, and Hypalon) that have wonderful wear-and-tear resistance but have poor electronic performance. These materials are primarily used in power cords, which operate only at 60 Hz, a very low frequency at which electronic performance doesn't matter. When higher frequencies are passed down wires, however, some effects begin to appear because of the plastic used.

Dielectric Constant and Capacitance

Frequency-based effects can be predicted by knowing one of two parameters, the dielectric constant of the plastic used or the capacitance of the finished construction. Every material used to insulate a wire has a dielectric constant, and two conductors have a capacitance between them.

The *dielectric constant* is a number that describes the quality of a non-conductor. The dielectric constant of a vacuum is defined as 1. The dielectric constant of air is very close to 1. The dielectric constant of any other material is greater than that (see Appendix A, "Notes and Comments," item 6).

The lower the dielectric constant, the better the plastic; the less energy stored, the more energy gets to its destination. That is why you often hear of plastics that cover wire described as dielectrics. If someone asks about the dielectric material, for instance, you know right away that they're talking about cables carrying information, data, or signals of some kind.

Capacitance is a measurement of how the cable design can "store" the electrical charge of the signal. It is based on two conductors and a material between them. You can change the capacitance of a pair of wires by

TABLE 2.9

Thermoplastic	Thermoset
PVC	Rubber
Polyethylene	EPDM
Polypropylene	Neoprene
Halar™	Hypalon™
Solef™	Silicone™
Teflon™	

Figure 2-7

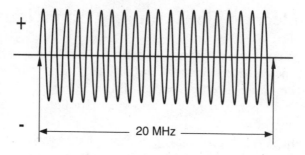

Figure 2-8

Going into a cable

\+

\-

20 MHz

spreading the wires apart, or adding more material between them, or by substituting a different material (Fig 2-7).

The unit of capacitance is the *farad*. Cable usually exhibits capacitance measured in billionths of a farad (picofarads, pF) between the conductors.

Capacitance is additive. In other words, a cable 1000 ft long has 1000 times as much capacitance as a 1-ft piece. The effect of capacitance worsens as the frequency of the signal on the cable goes up. Capacitance tends to store the signal, and the higher the frequency, the easier it is to store. Short runs of cable may be fine; problems may show up only in longer runs.

Figure 2-8 is a picture of a 20 MHz signal as it enters a cable. If the cable has a lot of internal capacitance, the next graph (Fig. 2-9) shows what the same signal might look like at the other end. If this were a 20 Hz signal, or even a 20 kHz signal, the outputs might look identical to the input. The difference lies in the materials used, the capacitance of the cable, and the length of the cable. Using better materials, you might pass 50 or 100 MHz, depending on the capacitance and the distance.

It is impractical in a twisted pair to spread wires apart unsupported to reduce capacitance. It is more logical to put more plastic on the wires to increase separation of conductors and yet allow them to still be twisted together. Higher grades of plastic (with a lower dielectric constant and, therefore, lower capacitance) also can be used. Putting more plastic on the wires makes them large and inflexible. Using fancy plastics keeps the overall size down but increases the cost.

Table 2.10 is a chart of various materials, the dielectric constants of those materials, and, in a "standard" configuration of a twisted-pair cable, the usual capacitance. Be aware that you could easily reduce (or increase) the capacitance as mentioned above. In a paired cable, intended for use as a signal-carrying cable, a manufacturer should always disclose the capacitance per unit length so that a comparison of performance can be made.

Note that the first three materials (rubber, neoprene, and EPDM) have very high dielectric constant, high capacitance, and are used almost exclusively in single-conductor wires, power cordage, or as jacket materials.

Inductance

Inductance is the electrical property of storing energy in the magnetic field that surrounds a wire. There must be electricity flowing to create this field. As soon as the electricity stops, the field will collapse and the

Figure 2-9

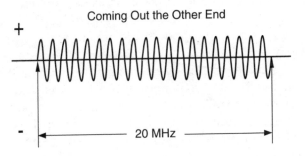

Coming Out the Other End

20 MHz

TABLE 2.10

Plastic Formulation	Dielectric Constant	Capacitance (pF/ft)
Rubber	3–6	50+
Neoprene™	3–8	50+
EPDM	2.5–6	50+
Polyvinyl Chloride	3.5–5.5	50+
Polyethylene	2.3	30
Polypropylene	2.23	30
Teflon™	2.1	30

energy will flow out of the wire. The amount of inductance in a straight piece of wire is very small, so its effect also is very small. We have to coil the wire up to get any significant inductance.

Time Delay

Some manufacturers prefer to indicate cable performance by citing the *time delay* of a cable. Time delay is usually measured in nanoseconds (billionths of a second) per foot. The faster the cable (shorter time delay), the better the performance. Time delay and dielectric constant are related to each other mathematically (see Appendix A, "Notes and Comments," item 6).

Coaxial Cable

Coaxial cable, or *coax* for short, consists of a single conductor, stranded or solid, surrounded by a dielectric. The dielectric is in turn surrounded by a braid or foil-and-braid shield and an outer jacket.

Coax is usually run with the shield at ground potential (i.e., 0 volts) and the center as the signal-carrying conductor. Because it is a single conductor shielded, it is no longer a balanced line. It is an unbalanced line. It is no longer a twisted pair, and it has poor crosstalk rejection. The shield is now one of the conductors in the circuit, as well as the shield, so its effectiveness is less than a separate shield over a twisted pair.

Impedance

When signals transfer from one piece of electronic equipment to another, especially at high frequencies, the cable must match the boxes that send and receive the signal. Impedance in the cable is determined by the size of the conductors, the distance between the two conductors, and the dielectric constant of the material between. Impedance is measured in ohms (symbol Ω), but do not confuse it with resistance, which also is measured in ohms (see Appendix A, "Notes and Comments," item 12).

The impedance of a coaxial cable is determined by the diameter of the center conductor, the diameter of the shield, the distance between the two conductors (center conductor and shield), and the dielectric constant of the material in between (Appendix A, item 23). The impedance of a cable is measured in ohms. Most common coaxial cable impedances are 50, 75, and 93Ω.

Once an impedance is selected, a cable of virtually any diameter can be made at that impedance. A coaxial cable's variation of impedance or its *impedance tolerance*—that is, variation in the relative position of the center conductor to the shield—is very low. This factor alone, even sacrificing all the advantages of twisted-pair and balanced lines, makes coax an excellent choice for video and other high-frequency uses. In fact, for very high frequencies it often is the only choice.

In twisted pairs, it is very difficult (though not impossible) to twist them together so that the impedance is constant. It usually varies widely. While the impedance of a cable is less important at audio frequencies, at video frequencies and beyond it becomes more of a problem. Making twisted-pair cable that supports video is difficult and expensive. But there is a much easier way to make a cable with a predictable impedance, that is, where the position of the wires is very predictable (Fig. 2-10). This is coaxial cable.

Attenuation

The key difference between big or small coax will be *attenuation*, which is signal loss on the cable. Attenuation can be caused by many characteristics of the cable; the resistance of the wire, the type of dielectric chosen, the capacitance of the construction, and variations in impedance are prominent factors. Attenuation is a key factor in comparing one cable design to another. The lower the attenuation, the better the cable.

Attenuation is expressed in *decibels* (dB) per unit length at a specified frequency. For example, a cable might have 6 dB loss per 100 ft at 10 MHz.

Figure 2-10

Cables can carry more than one frequency at a time. Cable television coax is carrying dozens (even hundreds) of different channels. The range of frequencies the cable is capable of carrying is called the *bandwidth*. Cable television will be using bandwidths up to 2 GHz (2 gigahertz, or 2 billion cycles per second) to carry over 150 channels.

Each channel is much smaller than the whole bandwidth. For instance, the entire FM radio band is 20 MHz wide. Each FM channel is only 200 kHz wide, however, and the actual audio sent is only two channels of 15 kHz each. Television, on the other hand, has a bandwidth (over the air) of over 400 MHz. Each channel is 6 MHz, of which the video signal itself is 4.2 MHz. So the cables necessary to carry a single video signal or the full TV bandwidth could be very different.

Velocity of Propagation

Dielectric constant is also related to another often-used unit of measurement, the *velocity of propagation* (V_p). This is a number related to the speed of light in a vacuum, which is defined as a V_p of 100 percent (see Appendix A, "Notes and Comments," item 6).

Air's V_p is very close to 100 percent. Most plastics have V_p of 66 percent or lower. To get closer to 100 percent V_p, you can foam the plastic to add air. This can get the dielectric up to 78 percent V_p.

The best dielectric would be nothing but air (or, ideally, a vacuum) inside the cable. But if the design used only air, how could the center conductor be held exactly in the middle, where it must be to maintain impedance? Cable designers have a constant battle between impedance tolerance and velocity of propagation. Figure 2-11 is a microscopic view of a foamed plastic.

Figure 2-11

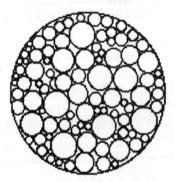

You can look at foamed compounds under a microscope and see their structure. In Fig. 2-11, the bubbles are of different sizes. This means there was poor control over the chemical mixture used to make the foam. It would be difficult to predict the velocity of propagation for this foam, since it would be difficult to predict what percentage is air and what percentage is plastic. Moreover, even if you could measure the velocity at a certain point, it is likely to be a different value somewhere else. So the key failing with this particular example is that the dielectric constant (and therefore velocity of propagation) would vary. Chemical foam also leaves a residue in the mixture, which affects the dielectric constant and cannot be removed.

Since the dielectric constant of this foam varies, if you extruded it around a conductor, the conductor's impedance also would vary. Variations in impedance cause *standing waves* or *structural return* loss (SRL). These phenomena reflect the signal back to the source and, in long cable runs, can be so severe that at high frequencies there could be virtually no signal at the end of the cable.

Velocity and Dielectic Constant

Here is the formula that shows the relationship between dielectric constant (DC) and velocity of propagation:

$$V_p = \frac{100}{\sqrt{DC}}$$

Or to convert from velocity of propagation (in percent) to dielectric constant:

$$DC = \frac{10,000}{V_p^2}$$

Table 2.11 lists some plastics and their velocities of propagation. Note that some velocities are slow (smaller numbers) and some are faster (larger numbers). The higher the velocity of propagation, the better the performance of this material at high frequencies. We can increase the V_p by changing from one plastic to another, or by adding air into the plastic (i.e., making a foam). Also shown are the dielectric constant and V_p of the same plastic when it is foamed.

TABLE 2.11

Plastic	DC	V_p	Foamed DC	Foamed V_p
PVC	3.5-5.5	45-57%	Too Soft	N/A
PE	2.29	66	1.64	78%
PP	2.25	67	1.64	78%
Teflon™	2.1	69	1.64	78%

Polyvinyl chloride is PVC, polyethylene is PE, polypropylene is PP. Then there's Teflon and some of its variations: fluorinated ethylene-propylene (FEP) and tetrafluoroethylene (TFE).

When it's foamed, the electrical advantage of dielectric constant (and therefore capacitance) of one plastic over another is lost. This is because much of the dielectric is now air and the plastic itself contributes much less to the electrical properties. These materials all have different dissipation factors, however, which may affect overall attenuation. The result is that each cell in a cellular foam is full of air, or nitrogen (the major constituent of air).

If we wish to make a high-quality foam, we must try to make all the bubbles the same size, avoiding the problem in our first foam example (Fig 2-11). By keeping the bubbles the same size, we will have a predictable capacitance, a predictable dielectric constant, a predictable velocity of propagation, and a predictable impedance. Figure 2-12 shows what that foam might look like.

Center Conductor Migration

The problem with big bubbles, when extruded over a wire, is that the foam is naturally soft. Conductors don't stay where they are supposed to be; they migrate vertically through foam.

This migration usually occurs when a cable is flexed while being installed. Solid wires have a greater tendency to migrate than stranded wires, because the stresses in them cannot be equalized among a group of conductors. The foam will allow the center conductor (in a coaxial cable, for example) to migrate, thus changing the impedance.

Crush resistance of a large-cell foam also is not good. If you step on this cable, it will change the impedance at that point.

Solid wires give better performance, however, because the shape (and therefore the position of the conductor) is more predictable, and at higher frequencies the skin effect is much better on a solid conductor than a stranded one. So for very high performance, solid conductors are always superior; some other method to overcome the disadvantages of foam is required. What we need is high velocity but hard foam cells!

Gas-Injected Foam

There is a cutting-edge method of making foam with a very high velocity of propagation. It involves injecting nitrogen gas into molten plastic (instead of chemically foaming, as is done to regular foam). Gas injection has a number of advantages. It is highly predictable, giving very constant results. This consistency results in very predictable capacitance, dielectric constant, and, therefore, very predictable impedance.

Because the bubbles are small, there are a lot of them per unit of area. You actually can get more air into these smaller bubbles, thus getting high velocities (80—85 percent or more). Figure 2-13 shows what a gas-injected foam might look like, compared to the preceding examples.

Hard-Cell High-Density Foam

By choosing the appropriate type of plastic, you also can get these small bubbles to have very hard cell walls. These will resist conductor migration and crushing. This will allow you to keep a reasonable impedance

Figure 2-12

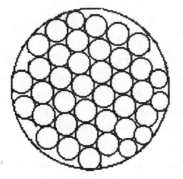

Figure 2-13

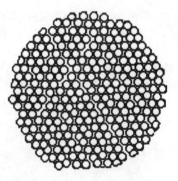

tolerance of ±3Ω, where standard foamed cables have a hard time keeping a tolerance of ±5Ω. This is called *hard-cell* or *high-density* foam.

Too Much of a Good Thing

High-density foam actually can get too good. The foam can get so hard and resistant to crushing that it has other problems. First is flexibility. Especially in larger sizes, high-density foam can result in very stiff cable—perhaps not as stiff as a solid dielectric, but certainly not what one would characterize as "flexible." Small-diameter cables would be more flexible, but at a compromise of attenuation, because smaller cables (all other things being equal) have poorer attenuation.

The second problem with high-density dielectrics is *center conductor adhesion*. One would expect this to be a problem with Teflon because nothing sticks to it—any Teflon frying pan will prove this to be true—but it is almost the same with high-density foamed polyethylene, the other most common dielectric.

The effect is to have horizontal conductor migration. This is sometimes seen after cables have been put into racks and cut to length. If connectors have not yet been attached, cable with center conductor adhesion problems will exhibit an odd effect: The center conductor slowly moves out from the cable until it is poking out the end and must be trimmed off. Often it can protrude half an inch or more. The effect can take place over a few minutes, or it may take a number of hours to show itself.

If connectors have been installed, the force of this migration can be so great that, for example, it can push the center pin completely out of a BNC video connector. In less severe cases, you may note that the BNC pin is at an odd angle. This is even more probable with digital video

BNC connectors, which give little or no support to the center pin. If one happened to cut off the center conductor at some angle other than 90°, the resulting pressure will push the pin off at an angle, making mating of that connector difficult or impossible.

Conductor Forces

When you take wire off a reel and bend it to a different shape, you are exerting forces inside the cable, especially on the center conductor (Fig. 2-14). Most multiconductor and coaxial cables respond to this force by trying to reduce the tension. They will migrate vertically, as shown in the lower part of the figure.

While this may be a minor problem with multiconductor cables, vertical migration can be a serious problem with coax cable. At the point of migration, the cable will no longer be its designed impedance and structural return losses due to impedance variation can be significant, especially at high frequencies or data rates.

Figure 2-14

Straight Cable and Center Conductor

Bent Cable and Center Conductor

Center Conductor Response to Bending

Figure 2-15

Horizontal Center Conductor Migration

Instead, we use high-density foam. This will resist vertical migration, which is excellent for impedance tolerance. But those forces generated by bending are still there. Where do they go? Since the center conductor cannot move vertically, it moves horizontally (Fig. 2-15), the only other way it can!

There are a number of techniques used to prevent horizontal migration. Among them are the following:

- **Pre-heating the conductor** This technique runs the conductor through a gas flame before it has the gas-injected dielectric extruded over it. As the heated conductor and melted plastic cool, the preheating helps them join together.

- **Gluing the conductor** The center conductor can be run through a liquid or powder glue solution. In the case of powder, it will melt when the hot plastic is extruded over it. The amount of glue attached is, of course, critical to how well the center conductor stays attached to the dielectric.

- **Preskinning** The center conductor is first run through an extruder, where a thin layer of solid plastic is put on the wire. This 'skin' will help a further foam extrusion to adhere.

Different Center Conductors

Stranded constructions have greater impedance variations because they are not always predictable in their position or shape. There is a compromise design that redraws the stranded material through a die, creating a *compacted conductor*. In maintaining flexibility while retaining some dimensional characteristics of a solid wire, this technique can be effective in some applications.

It is important that wires stay a certain distance from one another to maintain impedance and capacitance. You may have to sacrifice the high V_p of foam dielectrics for mechanical stability and low migration of solid dielectrics. But if you make a very stiff foam (to reduce the migration), a cable will have poor flexibility. The trick is to make lots of small bubbles, instead of a few big ones, and to make them hard-cell bubbles to reduce center conductor migration.

Hard Line

Hard-line transmission cable (Fig. 2-16) can get close to 100 percent V_p. (A value of 99.8 percent V_p is claimed). It is made by placing a polyethylene or Teflon spacer at intervals between the center conductor and the shield. This leaves mostly air in the cable. Hard line is as stiff as a pipe, however, because it *is* a solid copper pipe! It is very expensive, very heavy, and must be cut with a hacksaw.

Semirigid Line with Helical Spacer

One manufacturer, Andrew, makes a cable they call Heliax™ instead of coax. It is not as rigid as hard line and can be bent somewhat. The spacer inside holding the center conductor is a helix (spiral) made of Teflon or polyethylene (Fig. 2-17). There also are smaller, more flexible coaxial cables that use a helical spacer of polyethylene or Teflon with varying degrees of success. Smaller still are foam-filled versions. Other high-power transmission line manufacturers make similar products.

Figure 2-16

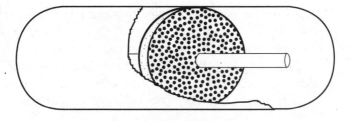

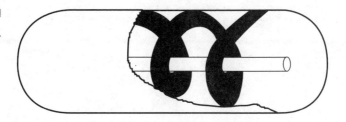
Figure 2-17

Time Delay

Time delay is the time it takes for a signal to travel down a cable, measured in nanoseconds per foot. The faster the signal travels, the less the time delay and the better the cable is. Time delay is directly related to the dielectric constant (and therefore, velocity of propagation). Here are some formulas. D_N is the delay in nanoseconds per foot, DC is the dielectric constant, and V_p is the velocity of propagation (see Appendix A, "Notes and Comments," item 6).

$$D_N = \sqrt{DC}$$

Reverse this to convert time delay to dielectric constant:

$$DC = D_N{}^2$$

Or the reverse to convert time delay to velocity of propagation:

$$V_p = \frac{100}{D_N}$$

Thus the delay of various dielectrics can be easily calculated. Table 2.12 shows some sample calculation results.

Group Delay

Most often it is not a single frequency you send down a cable, but a group of frequencies ranging from low to high. In audio, music is a complex signal of many frequencies traveling simultaneously down the cable. In video, picture information can be sent a number of ways. One is to split the picture information into its red, green, and blue constituents. The other way is to have a main signal, carrying the

TABLE 2.12

Dielectric Constant	Velocity of Propagation	Time Delay (ns/ft)
2.3	66%	1.52
1.64	78%	1.28
1.54	83%	1.24

black-and-white information, and a second signal (called a *subcarrier*) carrying the color information. Thus the picture information is spread over a range of frequencies.

If a cable varies in delay, impedance, or dielectric constant (and therefore V_p), parts of the signal can be affected differently (i.e., the high-frequency portion of the signal may arrive later than the low-frequency portion). This is called *group delay.*

Group delay is not a factor in audio. Variations in impedance, velocity, and delay are of little importance even at the highest audio frequency (20 kHz). If the group delay in a video signal (4.2 MHz) is very bad, however, the signal would be useless unless there were some reassembling of the picture elements.

Dielectric Absorption

A perfect capacitor discharges the same amount of electricity with which it was charged. In the real world, such as a cable, some of the capacitive charge is stored and not discharged.

This absorbed charge can affect fast pulses or waveforms with fast rise times, such as high-frequency signals with very short wavelengths. In a bad dielectric, this absorption can stretch a pulse or produce multiple false pulses as the discharge is delayed after the original signal has gone by. The effect correlates with dielectric constant.

Piezoelectric Noise and Dissipation Factor

There are other effects that make dielectric materials differ from each other. In the case of polyethylene (PE), polypropylene (PP), and polyvinyl chloride (PVC), their molecular structures are very different, even though (in the case of PE and PP) their dielectric constants are virtually identical. PP has a crystalline structure. PE has a much less defined structure. PVC is amorphous and has no crystalline structure at all. Because of this, PP has a *piezoelectric effect.* When moved or struck, the crystalline structure generates noise by itself, also called "self-noise."

At higher frequencies, the structural differences of dielectrics make attenuation (signal loss) and other parameters somewhat worse. This is why you never see polypropylene in, say, a CATV coaxial cable that must operate at very high frequencies.

PVC, while electrically "quiet," has such bad dielectric properties that it is used only at very low frequencies or where the electronic performance of the cable is not an issue.

Solderability and Shrink-Back

To technicians who turn raw cable into assemblies, solderability and shrink-back are major factors. In the process of soldering, heat is applied to the wire to melt solder and make a permanent connection. Many plastics deform or even melt off the wire under these conditions. Other plastics shrink back, leaving more bare wire exposed than was stripped originally. While different formulations can vary, and the actual temperature applied is a significant factor, Table 2.13 shows generally what to expect from the most common plastics.

Foamed versions of these plastics are worse still, and virtually all foamed cables are much better for termination procedures that do not involve heat (such as crimping, clamping, or insulation displacement).

Flexibility and Flex Life

Flex life is the measure of a wire's ability to be bent a number of times before it breaks. This is different from *flexibility*, which is a subjective measurement of "limpness." There are standard tests to determine flex life, but there are currently no standard tests to determine flexibility. In any flex-life testing, the conductors will almost always fail before any insulation does.

If there are two wires of the same gage, the wire that has the greater number of strands, or more twists-per-inch, will have higher flexibility and flex life than the lesser number of strands or fewer twists-per-inch.

Some insulation materials are more or less flexible than others. Table 2.14 compares flexibility to electronic performance.

Rubber, neoprene, and EPDM are often used on power cords. At the low frequency (60 hertz) where they operate, electrical performance is not an issue. At high frequencies, you might use something better (such as polyethylene) inside a cable and something more flexible as a jacket (such as PVC) over that dielectric.

TABLE 2.13

Plastic	Shrink-Back	Solderability
Polyvinyl Chloride	Good	Very Good
Polyethylene	Poor	Good
Polypropylene	Good	Good
Teflon	Excellent	Excellent

TABLE 2.14

Material	Flexibility	Electronic	Cost
Rubber	Excellent	Good	High
Neoprene	Excellent	Bad	Avg+
EPDM	Excellent	Bad	Avg+
PVC	Very Good	Avg	Low-Avg
Polyethylene	Poor	Excellent	Avg
Polypropylene	Poor	Good	Avg
Teflon	Bad	Excellent	Very High

One common, super-flexible jacket material is Hypalon. Hypalon is a CFC-based chemical construction which, being environmentally undesirable, will be phased out with all other CFC use. Other new chemical constructions, including a "CFC-free Hypalon," are now entering the marketplace.

Fire Ratings

The National Fire Protection Association (NFPA) publishes a book called *The National Electrical Code®* (NEC). It describes in great detail the safety aspects of wiring installation. Some sections apply to the flammability of wire and cable (see Appendix A, item 3). These guidelines put wire and cable (and fiber optics) in four basic groups: residential, commercial, riser, and plenum. The two issues the code addresses are the ability of a cable to support a fire, and the smoke produced during burning.

All cable will burn, given enough heat, but some will be better under certain conditions at stopping burning once the heat source has been removed (see Appendix A, item 4). Virtually all audio and video cables are based on five groups, unrated, residential, commercial, riser, and plenum

UNRATED cables are those not tested to the NEC specification. A significant number of older styles of cables are unrated. You should be especially cautious before doing an installation with unrated cable. Usually such cable has to be in conduit in order to meet code. You should check with an architect, fire inspector, or other responsible party before using it.

RESIDENTIAL rated cable is for use only in residential buildings. Cables must be less than a 0.250-in diameter.

COMMERCIAL is for commercial buildings. Cable can go through a wall without a conduit, but it cannot be run vertically between floors, or in drop ceilings or raised floors that are part of an air recycling system.

RISER is rated for any of the uses above, and also for vertical shafts between floors in a commercial installation.

PLENUM cable can be used in any of the areas noted above, including plenum areas. A *plenum* is defined as any hidden area (such as a drop ceiling or raised floor) where that area also serves as the return-air path for building air conditioning. Be aware, however, that many inspectors call all drop ceilings or raised floors plenum, even if they don't handle the air for air conditioning.

CL2, CL3, and CM

Within the classes above are voltage ratings on certain classes of cables. Class 2 cables (CL2) are "power-limited" but have no specific voltage rating. Class 3 (CL3) cables are also power-limited but are tested up to 300 volts. CM is a higher rating than either CL2 or CL3. Table 2.15 presents the four flammability ratings and three electrical ratings in a chart.

These twelve NEC designations are printed on cable to identify electrical and flammability rating. It is impossible to print on Teflon, so a tape is wrapped around the inner core with the cable identification and NEC designation. It can be read through a clear Teflon jacket. If there is no printing or labeling on any cable, it must be considered unrated.

TABLE 2.15

	Residential	Commercial	Riser	Plenum
Class 2 cables	CL2X	CL2	CL2R	CL2P
Class 3 cables	CL3X	CL3	CL3R	CL3P
Communications	CMX	CM	CMR	CMP

NEC Substitution

The residential class (X) is a standalone class. In other words, one may not substitute anything else for it. Table 2.16 shows the substitution cross list among the other codes.

Unrated cable is lower than anything on the chart. Any rating is better than unrated. Note also that the chart starts with the three electrical ratings, then repeats with a higher flammability rating for all three, and repeats this through the plenum flammability ratings. If a certain grade is required for a certain installation, no lower grade can be substituted. If a higher grade was chosen simply because of availability or familiarity, however, it is often possible that a lesser grade will meet the requirements of that installation. Here again, an architect, fire inspector, permit board, or other influence should be able to determine the minimum rating necessary to meet their requirements.

There are other grades of cables, such as MP (multipurpose) and FP (fire protection). These can substitute for some of the ratings listed above, but they are very rare or nonexistent in audio and video cables. The CL2, CL3, and CM families account for virtually 100 percent of all cables you might use.

Conduit

Any cable will meet any code requirement if it is in metal conduit. The cost of having conduit installed, however, can be much more than the added cost of plenum cable. In some cases even plenum cables are not an option. Some communities require that everything must be in conduit. Just be sure and check with the local inspector, fire marshall, or planning commission before finalizing plans. There is more on conduit in Chap. 11.

TABLE 2.16

CL2	CL3	CM	CL2R	CL3R	CMR	CL2P	CL3P	CMP
CL3	CM	CMR	CL3R	CMR	CMP	CL3P	CMP	
CM	CL3R	CMP	CMR	CL3P		CMP		
CL2R	CMR		CL2P	CMP				
CL3R	CL3P		CL3P					
CMR	CMP		CMP					
CL2P								
CL3P								
CMP								

Teflon vs. Plenum

When the NEC first defined a plenum cable, there was only one material that met the flammability specification and that was Teflon. Because of the expense and great difficulty of working with Teflon, many alternatives to it have been produced since then, including Solef™, Halar™, and Flamarrest™, as well as variations on Teflon itself, such as TFE or FEP.

Flamarrest is a PVC formulation that can meet plenum specifications. Since it has the same general electrical characteristics as regular PVC, you most often see another material as the inner dielectric (such as FEP Teflon) and Flamarrest as the outer jacket material. Still, it represents a significant cost savings, since PVC (even in special formulation) is much cheaper, has much greater flexibility, and makes cable preparation much easier.

Many engineers, designers, and inspectors still equate Teflon with plenum. It sometimes can be a laborious exercise to convince them that anything that says CMP, CL2P, or CL3P is rated for plenum, regardless of what material they are made. In other words, Teflon can indeed be plenum-rated, but not all plenum-rated cable is Teflon.

The only time a substitution cannot be made is if the cable must be able to function at high temperatures. Flamarrest is PVC and will melt at the same 75°C as regular PVC, while certain kinds of Teflon will function beyond 200°C.

The Local Inspector

NEC guidelines are just that, guidelines (see Appendix A, "Notes and Comments," item 3). They are voluntary. Any county, city, or even in some cases, individual inspector might or might not adhere to the code. It is absolutely essential that the building inspector, fire marshall, fire inspector, permit board, or other authorized individual sign off on plans in regard to required wiring practices before anything is installed. The broadcast industry has horror stories of entire installations being ripped out because the wiring did not meet the code requirement!

And remember that if a city, county, or state adopts the NEC as its guidelines, they become the law and are no longer just voluntary guidelines.

Cable Performance

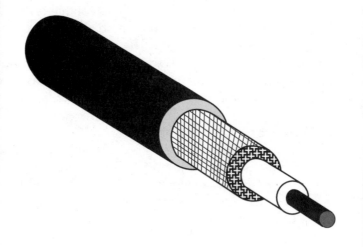

This chapter addresses some of the important electrical properties of cables that affect their performance, factors such as impedance, attenuation, and power- and voltage-handling capabilities. In addition, some of the ways these factors can interact are examined.

Impedance

Recall from Chap. 2 that impedance is the characteristic of a cable that works against the flow of electricity. In a twisted pair, impedance is calculated using the diameters of the two conductors in the construction, the distance between them, and the dielectric constant of the plastic used on the conductors.

In a coaxial cable, impedance is calculated using the diameter of the center conductor, the diameter of the shield, the distance between them, and the dielectric constant of the plastic used between. Large or small coaxial cable can be made with the same impedance by keeping the ratio of these factors the same.

Cables can be made in a number of common, "standard" impedances. Coaxial cable usually comes in 50, 75, and 93Ω impedances. Twisted-pair cable is most often seen in 100, 110, 124, and 150Ω flavors. No matter what the actual numbers, impedances are determined in the same general way (see Appendix A, "Notes and Comments," item 12).

Attenuation

All cables reduce the signal level as it passes down the cable. This is called *attenuation*. It can be caused by the resistance in the wire, the capacitance between conductors, impedance variations in the construction, inductance, and other factors. Attenuation is usually expressed as a loss of *x*-many dB (decibels) per unit length at a certain frequency. Thus cable X can be said to have lower loss at 50 MHz because it has an attenuation of 6 dB per 100 ft, while cable Y has 10 dB loss per 100 ft at the same frequency. Attenuation is also affected by the given impedance of a cable. In coaxial cable, a value of 77Ω gives the lowest attenuation. (See Fig. 3-1.)

Power Handling

With many types of cable, especially large-size coaxial cable, some users will wish to carry large amounts of power, such as the output of a transmitter or other radio-frequency amplification device. The efficiency of a cable's power handling (measured in watts) will be determined by the impedance chosen for that cable. The maximum power handling occurs around 30Ω (see Fig. 3-1 and Appendix A, item 11).

Voltage

Voltage is what drives a signal down the cable. With especially high voltages (above 600 volts), the dielectric material in coaxial cable can be critical to avoid arcing between the center conductor and shield. The ideal high-voltage point occurs at around 60Ω. (See Fig. 3-1.)

Contrasting Parameters

During the 1920s and 1930s, it was found that certain cable impedances worked best for the transfer of energy. For example, in video signals (which are generally high-frequency but low in power), the minimum attenuation occurs at 77Ω. Constraints of cable construction, such as using standard available gage sizes and standard mixtures of dielectrics, made a 75Ω cable easier to build.

In the case of transmission lines, which carry high frequencies and high power, the optimum power transfer occurs at 30Ω, but the maximum voltage rating occurs at 60Ω. Therefore, a 50Ω construction was the compromise solution. The actual mathematical mean is 51.5Ω, and you can still find some Mil-spec transmission cables of that impedance or higher (52 or 53Ω). They would be slightly better in voltage rating specifications than a straight 50Ω cable, although their power rating would be slightly less. To use them, transmitter and antenna would have to be tuned to that impedance, not 50Ω. (See Appendix A, "Notes and Comments," item 7.)

Attenuation curves (dB loss at a certain length for various frequencies) are commonly used to figure the loss of a cable. This information can be valuable to an engineer trying to design a system with the least loss (Appendix A, item 9).

We can show the relationship between impedance, attenuation, power, and voltage with the graph presented in Fig. 3-1.

Dielectric Voltage Ratings

Many users apply high voltages to coaxial cables, usually as transmission lines to antennas. In those applications, the breakdown voltage strength of the dielectric is very important. Different materials have different

Figure 3-1

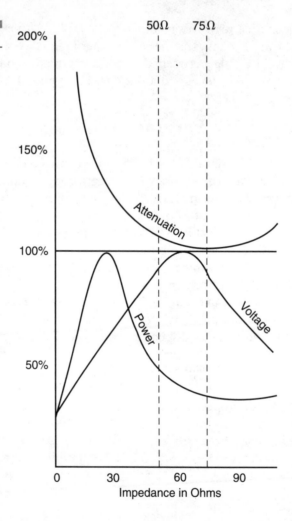

voltage ratings. In all cases, as more and more material is added, the lower the voltage rating of each successive layer. The ambient temperature and frequency being transmitted also affect the voltage rating of any construction.

Teflon is the material often used for high-voltage applications. This is not because of any special properties at high voltages, however. (In fact, polyethylene will outperform Teflon.) But high-voltage transmission lines can often get very hot, especially as connectors and joints age. Where resistance increases, heat is generated (Appendix A, item 11). Because Teflon has the highest temperature rating, it will last the longest before it melts and the line fails. This is also why a high ambient temperature can hasten failure, by helping the cable arrive at that melt point sooner.

Foam dielectric designs are not recommended in high-voltage applications because, at the melting point, foam breaks down rapidly. The chance of transmission line failure more than offsets the added velocity of propagation or reduced dielectric constant of foam. Solid dielectrics or air-core with solid support, are the best choice for high voltages.

Figure 3-2 is the voltage deration curve for Teflon. Note that the first few mils have tremendous voltage strength and that this derates rather quickly to a fairly standard value. (These values are measured in an oil bath to remove the influence of air, which severely derates the dielectric strength.)

If power is not being sent down a transmission line (if a signal is being received, for instance), the temperature and voltage considerations of any cable are of little consequence. Connectors and joints do affect overall attenuation, so their aging and reduced performance are still a factor.

Balanced and Unbalanced Lines

In Chap. 1 we discussed *balanced lines,* where the level of signal on each wire in a pair is the same at any point in time, but of opposite polarity. If the signal from each wire could be "added" algebraically, the total would be zero. (That is, with +1 volt on one wire and —1 volt on the other, the total is zero.)

Balanced lines allow significant reduction in electromagnetic noise because noise striking both conductors has the same polarity on each—unlike the signal on the cable, which has opposite polarities on the two

Figure 3-2

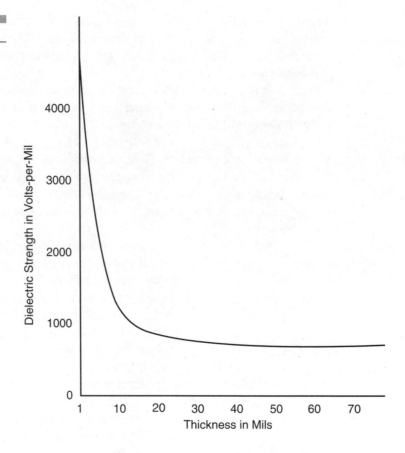

conductors. The noise will cancel itself out when the signal reaches its destination. This is called *common mode rejection* because the noise is in the same, or "common," mode, and is therefore rejected.

Twisting pairs helps avoid noise and unwanted signals from adjacent wires, a phenomenon called *crosstalk* in which one pair's signal will "cross" to another pair. It is not uncommon to hear some other conversation (sounding very distant) while on the telephone. That is crosstalk.

In coaxial cable there are two conductors. One is a single wire running down the center of the cable, and the other is the shield itself. Since the shield is at ground potential (i.e., 0 volts with respect to ground) and the signal runs down the center conductor, the two conductors are not balanced. They do not have equal and opposite polarity on them. They form an *unbalanced line*. Since it is not a balanced line, coaxial cable loses the advantage of the common mode noise rejection.

The vulnerability of a signal on a cable to interference also is based on the level of the signal. If it is a strong signal, it is more resistant to interference. This is one reason that video signals can be run unbalanced. They are typically at a level of 1 volt, which is much higher than microphone audio levels, for example, which at 1 microvolt run a million times smaller.

Cable Construction

Figure 4-1

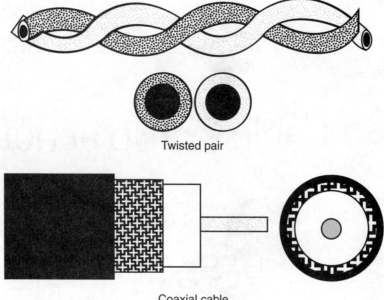

Twisted pair

Coaxial cable

Two cable constructions are commonly used to carry audio, video, or data signals. One is twisted pairs, the other is coaxial cable (called "coax"). Figure 4-1 shows cross-section pictures of each.

The twisted pair is two wires, with a dielectric extruded over each, that are twisted together. Coaxial cable is a single conductor with an extruded dielectric, covered by a conductive shield and outer jacket. We show solid conductors in the twisted pair and coax, but they could just as easily be stranded.

Shielding

Besides twisting wires, we can also protect the signal from interfering with adjacent pairs, or from some intrusive outside signal or noise, by shielding the pair (Fig 4-2).

There are three basic types of shields, a *serve* (also called a *spiral*), a *braid* and a *foil*. The serve and braid use individual conductors wrapped around the dielectric. The foil is aluminum foil wrapped around the dielectric. There are six considerations when choosing shielding:

- **Coverage** How much of the inner cable will be covered? This usually is expressed in percent.

- **Flexibility** How limp is the construction when using a certain style shield?

- **Flex life** How will the shield stand up if it is to be constantly flexed?

- **Frequency range** How does this shield work at low or high frequencies?

- **Triboelectric noise** When flexed, does the shield open and close, producing a change in capacitance in the cable that can be heard as electrical noise? Will it rub against the dielectric and build up static charges, which also can be heard?

- **Multipair crosstalk** If we put more than one shielded pair in a cable, will the shields touch each other, increasing crosstalk from one pair to the other?

The layout of a serve or spiral shield is shown in Fig. 4-3. A serve or braid shield is made of individual wires wound or braided around the central pair. It usually is made with copper or tinned copper wire; aluminum sometimes is used in inexpensive cables.

No matter how perfectly such a shield is made, there always will be gaps between successive windings. As the frequency of the signal on the

Figure 4-2

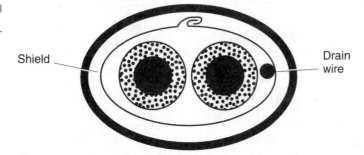

Shield — — Drain wire

Figure 4-3

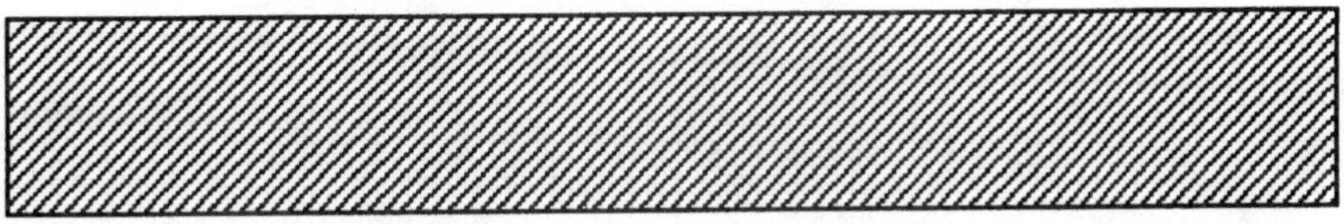

cable increases, the small slots or holes in the shield are closer and closer to the length of the wave of each cycle of the frequency. In other words, the shield becomes less and less effective at higher and higher frequencies; it is said to be *wavelength dependent*.

Because of this, there currently is a controversy regarding serve or spiral shields, especially with respect to high-frequency (analog video) cables. Most "hi-fi" cables are serve-shielded, and many claim that serve shields are only good to 30 kHz (that is, good for audio only). They claim the spiral wrap acts like an inductive coil and inherently limits the frequency range. This effect can be reduced somewhat by putting a reverse second serve shield over the first.

The inductive effect is why standard hi-fi hook-up cables are poor choices for video interconnection. On the other hand, most top-of-the-line home video recorders, or bottom-of-the-line professional video recorders such as Super-VHS (S-VHS), use serve-shielded cables up to 4 MHz!

While the frequency issue is debated, the coverage is well-known. Coverage can be excellent, as high as 95 percent. As you bend a serve shield, however, it tends to open up, reducing its crosstalk effectiveness and making it "noisy." You can reduce this effect somewhat by putting a reverse second serve shield over the first. This is called a *Reussen shield*, and it is very popular in Europe for its flexibility. Still, most engineers agree that the performance of a serve shield is inferior to a braid shield.

There is a new kind of braid shield called a *French braid* (Fig. 4-4). It is a combination of two serves, where groups of served conductors are alternated, or braided, with others, "locking" the two serves and preventing them from opening up when flexed, as a single serve or dual serve construction will.

French braid preserves the flexibility of a serve shield, with coverage and stability which approach that of a standard braid shield. Its effectiveness above audio frequencies is also excellent.

French braid is a patent of Belden Wire & Cable.

Braid shields (Fig. 4-5) are made by weaving bare or tinned copper wires over a core. If you put two layers of 95 percent braid on a cable, you can get up to 98 percent coverage.

Figure 4-4

Figure 4-5

As the frequencies inside the cable (or noise outside the cable) approach the size (wavelength) of the holes between groups of wires, braid shields become less and less effective (see Appendix A, "Notes and Comments," item 15). Shield coverage begins to be wavelength-sensitive at 10 MHz, and by 400 MHz braid shields have dramatically reduced performance.

It should be noted that braid shields are used in high-frequency transmission up to 1 GHz and beyond. In these cases the shield effectiveness is not at issue, because they are not effective in shielding. What is important is the quantity of copper, low resistive losses, and good impedance stability, all of which can be achieved with a braid shield.

By adding a second braid and changing the angle of its mesh pattern compared to the first braid, high-frequency shielding can be extended. The different braid angle has a different wavelength dependency, and because it is layer-upon-layer, it decreases the apparent size of the holes. Where single braids begin to be wavelength-sensitive at 10 MHz, a double-braid with a varied braid angle can move the wavelength sensitivity and shield effectiveness up to approximately 50 MHz.

Braid shielding is the slowest, most expensive, most labor-intensive part of any cable construction. Braided shields are thus a good indicator of cost. All other things being equal, a braided shield will indicate a more expensive cable than an unshielded or foil-shielded construction.

For truly effective high-frequency shielding, we must look at foil shields. These are applied by winding the foil around the cable as it passes by the winding machine (Fig 4-6). This is a very easy and cost-effective way to shield a cable, much cheaper than labor-intensive braiding.

Most foil shields are made of aluminum. If adjacent layers overlap, foil by its very nature provides 100 percent coverage, so there is no wavelength dependency. Foil itself is very thin (often 2 mils or less) and thus is very flexible. At the same time, it is very brittle with poor flex life, and after only a few flexes it will crack and open up, reducing shield effectiveness.

Figure 4-6

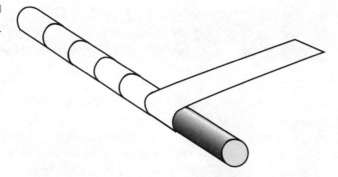

Figure 4-7

If you glue aluminum foil to a very flexible polypropylene or poly-ester film (such as Mylar Fig. 4-7) and then wrap it around the cable, you can greatly increase flex life.

If foil is merely wrapped around the cable, as Fig. 4-6 shows, flexing the cable will allow the foil to "open up," producing voids or unshielded areas. Note also that there is no foil-to-foil contact (Fig. 4-7). The cable is not completely shielded. This compromises shield effectiveness.

The foil can face either in or out of the core. If the foil faces in, the Mylar can be made in various colors, for instance to color-code each pair in a multipair cable. In multipair or multicoax designs, Mylar/foil great-ly reduces but does not eliminate contact from shield-to-shield crosstalk. A piece of foil from one pair can be accidentally pushed between layers of foil around the adjacent pair. If foils touch, no matter how slightly, crosstalk increases.

If this foil is applied over a paired cable, most likely there will be a drain wire on the conducting side (Fig. 4-2). This is the electrical connec-tion to the shield. The drain wire also allows ground continuity between the broken foil sections, but a drain wire cannot restore the shield effectiveness that was lost.

Whenever a foil shield can separate at the edges, as this type can, the changes in capacitance in the cable can be heard at the receiver. This is called *triboelectric noise* and can be a minor or major problem. If the cable moves, the shield can rub against the core dielectric and generate static noise, also a form of triboelectric noise. If the signal level in the cable is very low, even the slightest noise can disturb the signal purity.

On the other hand, if the cable is installed and never moved while in operation, you may never experience triboelectric noise. Triboelectric noise is a major issue in microphone cable and cable television drop cable (which goes from pole to house) because both of these can be in constant motion during operation.

One improvement is the use of bonded foil, where the foil is glued either to the core (common in coax cables) or to the jacket (common in paired cables). Bonding the foil greatly reduces triboelectric noise and helps prevent foil breakage.

In coax cable, there usually is a braid over the foil, which can join the broken sections and somewhat make up for (depending upon braid coverage) the lost shield effectiveness. Braid is most effective at lower frequencies, however, so the high-frequency shielding may be adversely and permanently affected by a broken foil.

By folding the polyester/aluminum foil along one edge, coverage is 100 percent (Fig. 4-8). Foil-to-foil contact is maintained. This design can still open up and cause voids in coverage, with no improvement in triboelectric noise. In multipair or multicoax constructions, stray pieces of foil from adjacent pairs or coaxes are now twice as likely to cause interference as in the preceding design, since there are now two pieces of foil in contact facing the outside.

Many inexpensive paired cables are made this way in the belief that the extruded jacket over the pairs will keep the pairs from moving around and the foils opening up and making contact. What is needed is a foil-to-foil design that won't open up.

With the double type of foil shield (Fig. 4-9), often called a "Z-fold," you have the ideal combination: a foil-to-foil "shorting fold" for 100 percent coverage, a polyester/foil combination for good flex life, and a folded design with an "isolating fold" that leaves no exposed foil, yielding excellent crosstalk performance and low triboelectric noise.

This shield is complicated to apply. The dual folding must be done while the cable is in motion and the polyester/foil tape is being wrapped around it. Very few manufacturers can do this. In multipair

Figure 4-8

Figure 4-9

TABLE 4.1

Type	Cost	Resistance	Flexibility	Flex Life	Frequency Performance	
					Low	High
Serve	Moderate	Excellent	Excellent	Excellent	Fair	Poor
French braid	Moderate	Excellent	Excellent	Excellent	Good	Fair
Braid	High	Excellent	Good	Very Good	Good	Fair
Foil	Low	Fair	Good	Fair	Fair	Good
Foil/Braid	High	Excellent	Good	Good	Good	Good

constructions, by asking a manufacturer if each foil shield is double or Z-folded, you can separate potentially noisy designs from the less-noisy.

Table 4.1 is a comparison chart of shielding effectiveness for some types of shields. Figure 4-10 shows the general difference in performance between braid and foil shields of all constructions at various frequencies.

Foil fails in one area: low-frequency signals. These signals contain a lot of energy. To keep these signals inside the cable (or to keep them from getting in) requires low resistance, i.e., a shield with large conductors—such as a braid shield. So while good at high frequencies, foil shields can only be rated fair at low frequencies. They lack the mass of a braid shield.

To get ideal shielding for both high and low frequencies, a foil-braid combination is used. This is normally braid-over-foil. The effectiveness of such a construction is still limited by the percentage of braid coverage, so the ideal combination is a minimum of 85 percent braid plus 100 percent foil.

To increase shielding effectiveness, especially in coaxial cable where there is no inherent common-mode rejection, multiple shields sometimes can be used. Double-braid shields can get up to 98 percent coverage. Double-foil tape (foil-polyester-foil), called Duofoil, also is used. It has many of the same properties of Z-fold shields but is much easier to apply.

A foil-braid-foil combination, called a "tri" shield, is common in cable television constructions, as is a foil-braid-foil-braid, or "quad" shield (Fig. 4-11).

Since a foil shield cannot be soldered to directly, a drain wire is provided in cables with only a foil shield. This is a tinned wire, sometimes the same gage as the conductors and sometimes smaller, that runs in contact with the foil. It allows one to make connection to the shield. It is never a bare (untinned) conductor because of the corrosion potential between copper and aluminum. (Tin and aluminum have much reduced potential.)

Transfer Impedance

Transfer impedance is a test of how much noise protection a shield will give a cable. In this test, a signal is introduced at one point on the outside of a shield. The center conductor of the cable is measured for signal

Figure 4-10

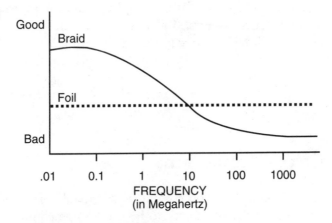

Figure 4-11

intensity. The lower the signal, the higher the transfer impedance and the better the shield effectiveness. If there is more signal on the center conductor, it means that signal must have leaked through the shield, meaning it is less effective.

Braid Shields and Bonded Foil

If the cable is for high-frequency use only, such as cable television (CATV), braid shields are ineffective. Braid shields are added only to facilitate connectorization. Braid gives the connector something to grab hold of when it is crimped or clamped in place. Thus CATV cable often has a foil/braid shield where the braid gives only 67 percent coverage, or 61 percent, or even as low as 40 percent. This does not indicate the shield effectiveness. At high frequencies, the only bad effect a reduced braid will have is shorter flex life (fewer strands) and lower connector reliability.

In high-frequency coax, because the shield is part of the impedance-specific construction, it is necessary to immobilize the shield to get good, predictable impedance. One way to do that is to glue or bond the foil to the dielectric. While slightly more expensive than loose foil, bonded foil construction greatly improves impedance stability and also increases flex life. For connectors, bonding also prevents the foil from being pushed back as the connector is being pushed on.

For the ultimate in broadband shielding, you can combine high-coverage braid with foil. Braid gives superior low-frequency shielding. Foil gives superior high-frequency shielding.

Coax Connections

To put a connector on coaxial cable, the cable is stripped to dimensions required by the connector to be used. Some connectors (BNC, for example) require a soldered or crimped pin on the center conductor. The pin is then inserted into a shell, which itself crimps or clamps on the shield. Other designs (notably the F connector) use the center conductor itself as the center connection. In the latter design, the center conductor must be solid, not stranded.

Triaxial Shielding

There is one type of cable that uses dual separated shields. This is called *triax* cable (Fig. 4-12). The most common use for this is connecting television cameras to other equipment in broadcast studios. In triax cable, the shields are separated from one another by a layer of plastic so that they each can carry signals or power to run the camera and equipment attached to the camera.

Triax cable should not be confused with dual-shield cables, in which two layers of shield are braided one on top of the other. The two shields in triax are physically and electrically separated.

All triax cables use braided shields because these shields are often used to carry power, so large conductors with low resistance are required. In addition, in the studio environment many triax cables are in motion with the camera while it is in operation. Both of these requirements preclude the possibility of foil shielding.

Hum, Noise, and Twisted Pairs

In twisted-pair cable, merely twisting the pairs gives significant electromagnetic interference (EMI) protection. At frequencies below 1000 Hz, in fact, it is the only protection you commonly have available.

The closer and more exact the spacing between the conductors, the more balanced they will be, resulting in greater noise reduction. Table 4.2 shows the noise reduction at various twist ratios.

Note, however, that it is not just the number of twists that improves noise-reduction qualities, but also the predictability of position that tight twisting imparts. The more predictable the spacing, the more balanced the pair and the greater the noise reduction. Just twisting wire

Figure 4-12

TABLE 4.2

Twist Ratio	Noise Reduction
Zero (parallel wires)	0 dB
3 twist/ft	23 dB
4 twist/ft	37 dB
6 twist/ft	41 dB
12 twist/ft	43 dB

together a lot may mean nothing for noise reduction unless such twisting results in more uniform conductor-to-conductor spacing.

In this regard, there is new technology called *bonded pairs*. These are conductors joined together during extrusion. They have extremely stable spacing and correspondingly high EMI rejection. They also exhibit extremely stable impedance, making such constructions excellent for high-frequency twisted-pair applications.

Loop Area

For analog audio or video, even double-braid shields are ineffective at frequencies below 1000 Hz. This is why 60 Hz interference from ac power wiring is so difficult to guard against, much less remove. The most important thing to consider with very low-frequency hum and noise pickup is *loop area*. This is the distance between the conductors in a balanced line, or the distance between center conductor and shield in an unbalanced line. In coax it is difficult to keep loop area small. Using tiny coax severely affects many other parameters, especially attenuation. There are certain techniques, however, that can be employed with twisted-pair balanced lines.

■ **Filled cables** These cables have insulating materials packed around the conductors to stabilize the impedance. Many older-style microphone cables have hemp (twine) or other fillers, but these are to keep the cable round and improve ruggedness. Digital-audio cables have polyethylene (or other) filler threads specifically to help stabilize impedance. "Filling" a cable this way also helps keep the cable smooth and round, which is aesthetically appealing. While not specifically designed to reduce loop area, fillers help prevent conductors in paired cables from wandering and separating inside the jacket.

- **Quad cable** Quad or star-quad microphone cable (not to be confused with quad-shield coaxial cable) uses four color-coded conductors spiraled together. Wires are combined to get to the two conductors needed for a balanced line, i.e. the wires opposite each other are combined into a single conductor, resulting in a twisted pair. Quad designs have extremely good hum and noise rejection, on the order of 30 dB more than standard two-conductor designs.

- **Bonded pairs** There are designs for data cables (such as Belden 1700A and 1872A) in which the conductors are joined or bonded together. The spacing is very small and predictable, with a very small loop area.

You usually are forced to use standard paired cable, and might still have to deal with hum and low-frequency noise problems. There are four further solutions to low-frequency interference:

- **Avoid it** Make sure the cable does not run along ac power wiring, extension cords, ac plug strips, etc. If ac and signal wires absolutely must cross, try to have them cross at right angles so the interaction is minimal. When dealing with audio and ac, use the inverse square law: Twice the distance produces one-quarter the effect. The farther apart they are, the better.

- **Hum-bucking** There are specialized transformers that pass low-frequency signals but trap induced hum—an expensive solution, especially if there are many lines. They are common with video in coax because of coax's inherent lack of protection from hum.

- **Mu-metal shields** These specialized conductors can be wound into shields for low frequencies, but they are extremely expensive and difficult to apply. No standard products use this material.

- **Armor** Any cable made can be armored at a moderate additional cost. A corrugated steel jacket is wound around the cable (usually with the addition of another jacket overall). Aluminum armor has little shielding effectiveness. Steel is the only effective material. At 60 Hz a corrugated 0.025-in steel armor increases shield effectiveness by approximately 18 dB.

- **Conduit** This is the most familiar solution, though cost is certainly a factor. Solid-steel conduit shield effectiveness at 60Hz is

TABLE 4.3

Material	Colorability	Ruggedness	Flammability	Printability	Cost
PVC	Excellent	Good	Fair	Excellent	Low-Medium
Polyethylene	Good	Very-Good	Bad	Good	Medium
Polypropylene	Good	Very Good	Bad	Good	Medium
Flamarrest™	Good	Fair	Excellent	Excellent	Medium+
Teflon™	Fair	Excellent	Excellent	Bad	High

20 to 27 dB, depending on the thickness of the conduit and the workmanship at the joints. Flexible steel conduit is about 18 dB effective. What is needed is massive conductors, which is why the thickness of the conduit is important. This is also why most wrapped shields (even Zetabon and other steel foils) have little effective shielding at 60Hz. They are just too thin; they lack the required mass.

Jackets

Jacket materials cover wire constructions. Ideally, you choose a jacket material based on mechanical properties such as ruggedness, weatherability, flammability, colorability, printability, etc. At higher frequencies, jacket material can affect cable performance. Table 4.3 lists some mechanical properties of jackets.

A great deal of effort has been made to produce cables that will meet plenum specifications and, therefore, save the considerable expense of installing conduit. One of these is Teflon. Teflon's main properties are its ability to withstand high heat, not support a fire, and produce very little smoke. The last two of those are specified in the NEC rules. But Teflon is very expensive, difficult to work with, difficult to color, label, or mark... in other words, a pain!

Things have changed in the last couple of decades, however. There are other materials that can meet plenum specifications. Some, such as Halar and Solef, are very similar to Teflon. There also are PVC compounds (such as Flamarrest by Belden) that can meet plenum specifications in some constructions. This does not mean that those constructions can withstand high heat; you still end up with Teflon for those. But Flamarrest constructions will meet the NEC requirements of low smoke and not supporting a fire. There are significant cost savings and much easier stripping and connectorizing, compared to Teflon.

PVC compounds like Flamarrest have only fair electrical properties, very similar to regular PVC. Where good electronic performance is required, more often you will see Teflon used as a dielectric inside the cable and Flamarrest as a jacket.

Weathering is a major consideration for cable that is to be used outdoors. In those cases, special jacket compounds can be used to reduce sunlight and ozone susceptibility. Other compounds, such as high-density polyethylene, can be used to make the jacket super-rugged for direct burial.

For direct and continuous water immersion, every plastic eventually will fail by absorbing water and passing it to the layers underneath. In those cases, solid armor and gel-filled layers are required to get a reasonable amount of life out of any construction.

Compound Migration and Contamination

Compound migration is a key element in the choice of jacket materials. Not to be confused with horizontal or vertical center conductor migration, *compound migration* is the effect of jacket materials on other materials. For instance, if a cable made with jacket compound X is laid on a piece of equipment made with plastic Y, various chemical reactions might occur. These range from discoloration to brittling to dissolving, depending on the two compounds. Most cables made today use nonmigrating compounds formulated to be inert and nonreactive.

Contaminating compounds in the jacket also can affect the dielectric underneath. They can discolor the dielectric and, in the worst case, change the capacitance, dissipation factor, and dielectric constant. Changing these parameters therefore can alter the cable's impedance, velocity of propagation, voltage handling, frequency response, and many other parameters.

Most cable manufactured today uses noncontaminating compounds that either do not migrate or, if they do contaminate, have no effect on the electronic parameters of the cable.

A significant number of cables made offshore suffer from migration and contamination problems, signs of poor quality materials, and poor manufacturing practices. Many of the compound migration problems are worsened by outdoor exposure. When used in indoor installation, the effects may not show up for months or years, long after the cable is installed.

In cables with foil shields, or foil-braid shields, the foil helps to prevent migration from the jacket to the dielectric; in those constructions it rarely is a problem.

Temperature Considerations

Plastics all react to temperature, but each plastic will react differently. Jackets are especially vulnerable because they are always exposed and are the first line of defense. For extremely low temperatures (below —20°F), most plastics fail, often becoming brittle and cracking. Even if subsequently warmed, many materials stored at low temperatures, especially PVC, may remain brittle and crack.

Exotic plastics, such as polyurethane and silicone, can survive low temperatures and still retain flexibility—but at a dramatic increase in cost.

Audio

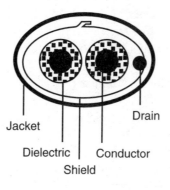

Jacket

Drain

Dielectric

Conductor

Shield

Analog Audio

An *analog* is "a thing that is like something else." Analog signals are copies of other kinds of signals. For instance, Fig. 5-1 is a diagram that shows the compression and expansion of air pressure caused by a single tone from a sound-producing device such as a musical instrument or human voice.

The air pressure goes up and down, forced by the pressure from the musical instrument (such as a violin or guitar string), or from the vocal cords in the throat of a person speaking. If you looked at the eardrum of a person hearing the sound, its movement would be an analog or copy of that series of vibrations. In other words, the eardrum would be vibrating just like the original sound. Although the figure shows a wave every inch or so, this is just a representation. As mentioned in Chap. 2, acoustic (sound) waves are much shorter than electronic wavelengths at the same frequency. The acoustic wavelength at 20 kHz is 0.676 inch. The electronic wavelength of 20 kHz is over 9 miles!

If this sound were picked up by a microphone, the diaphragm in the microphone would also vibrate in an analog of the original sound. Coming out of the wires from that microphone, generated by the coil-and-magnet arrangement, would be the electronic analog of the original sound. Figure 5-2 is the analog audio signal.

Audio generally occupies the range from 20 Hz to 20 kHz. Compared to radio frequencies or data bandwidths, these are low-frequency signals. When signals get to much higher frequencies, and the wavelength gets shorter, eventually the spacing of the wires, the spacing of braid shields, the dimensions of the cable, and even the spacing of the pins in connectors, get closer to the wavelength and begin to have an effect on the signal.

Because analog audio cables are low-frequency constructions, and the wavelengths are very long, the impedance of the cable is of no conse-

Figure 5-1

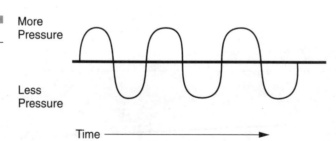

More
Pressure

Less
Pressure

Time ⟶

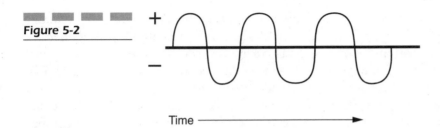

Figure 5-2

Time ⟶

TABLE 5.1

Capacitance between conductors (pF/ft)	Highest frequency (kHz)			
	15	20	30	40
12	1474	1111	737	556
19	931	701	465	351
24	737	555	368	278
31	571	430	285	215
50	354	267	177	134

−3dB (half-power) length in feet

quence. The dimensions of the cable are very different from the wavelengths of the signals it carries. Impedance variation throughout a run of cable is equally unimportant. The only exception is the phone company, which often has lines that are miles long. They then must compensate for losses in the lines by adding filters and equalizers. But miles-long runs are almost unheard-of in a standard audio installation.

The capacitance of various dielectrics, however, can cause signal degradation at audio frequencies; this can be predicted and measured. Table 5.1 is a chart that shows the half-power point (3 dB loss) on a piece of cable, as determined by the highest frequency traveling down that cable, with common values of capacitance per foot and a standard 600Ω load impedance. A different load impedance would affect these distances. For a 1-dB loss, divide the length by 3.

You also should be aware that, if the cable is made unbalanced by attaching one of the twisted-pair wires to ground, the capacitance of the cable will be approximately doubled, because the shield has been added as a signal conductor. This greatly reduces performance over distance.

One other factor in distance is wire gage. While the industry standard is 22 AWG, many installers and designers are moving to 24 AWG because of the cost and size savings. There are two compromises in such a choice. The first compromise is ruggedness. Just because it is more wire, 22 AWG will last longer when moved, flexed, or abused. If it is used in an installation, however, such properties are of little consequence; the reduced size, weight, and cost of 24 AWG can have added benefit. The second compromise is an increase in resistance (Table 5.2).

Electrically a cable can be characterized as a series of resistors and parallel capacitors, bringing into play what is called an *R-C time constant*. The R-C time constant is what ultimately affects high-frequency performance and creates "roll off" of high frequencies. See more on R-C time constants in the "Digital Audio" section later in this chapter.

Because of the small size and reduced cost, foil shields are very common in all-analog, line-level audio applications. While foil shields lack the low-frequency effectiveness of braid shields, they are much easier to make, much cheaper, more flexible (although flex life is not as good), and much smaller in size than braid-shield cables. Multipair cables with

TABLE 5.2

Resistance Per Thousand Feet (Ohms)		
AWG	Resistance (solid)	Resistance (stranded)
30	103.2	112
28	64.9	70.7
26	40.81	44.4
24	25.67	27.7
22	16.14	17.5
20	10.15	10.9
18	6.39	6.92
16	4.02	4.35
14	2.53	2.73
12	1.59	1.71
10	1	1.08

individually foil-shielded pairs (also called *snake cables*) sometimes include an overall shield for additional shielding protection.

Microphone Cables

In microphone-level applications (usually 50—60 dB below line level), losses, noise, and interference become more critical. For permanent, non-moving installation, the same foil pairs can be used. You must be more aware of the sources of interference, however, because foil shields are less effective at low frequencies.

Interference can be constant, such as from power wiring or high-voltage power lines, ballasts in fluorescent lighting, or permanent transmitters. Interference also can be transitory, such as from passing cars or devices with motors that cycle on and off (such as refrigerators or air conditioning). Still other sources can be completely random: ship radar, CB, cellular phones, 2-way radio, or other moving transmitter sources. Avoid running microphone lines near or parallel to anything that may generate an electromagnetic field. In many cases, solid-steel conduit can be a major help in reducing induced noise in microphone lines.

Microphone or "mic" cables are usually made for flexibility, flex life, and low handling noise, and not necessarily for performance. The best dielectric materials are somewhat inflexible, so most microphone cables use rubber, neoprene, PVC, or similar materials. While mic cables are excellent from the microphone to the connector in the wall or stage box, they are not intended for long runs. Run foil pairs (either single or snake) from that point.

Flexibility and flex life also can be affected by the metals used, the strandings used, and the plastics used to jacket them.

While copper is generally used, cadmium bronze is physically stronger and has much better flex life. Cadmium, however, is toxic when vaporized and is difficult to work. Copper alloys now being developed approach the strength and flex life of cadmium bronze, without the cadmium but at added cost.

All other things being equal, the higher the strand count, the greater the flex life. Thus, microphone cable designs with each conductor having 19 strands, 40 strands, or even 105 strands are readily available, each with increased flex life.

Microphone jacket materials include rubber and its artificial cousins (EPDM, neoprene, Hypalon). They are very rugged and have good flexibility. PVC, especially some of the new matte finish versions, can come very close in flexibility, though not in ruggedness. Other materials, such as polyethylene or Teflon, can be very rugged but are poor in the flexibility department. Since flexibility is of major importance in microphone cable, these last materials are rarely used.

Quad Microphone Cables Quad or "star-quad" microphone cables (Fig. 5-3) are four-conductor microphone constructions that have very low noise and hum pickup. The four conductors are wound together in a spiral. To get the added noise rejection, join the wires that are across from each other into a single pair. Usually these are color-coded to aid in identifying which to combine.

Using Fig. 5-3 as a guide, after stripping off the jacket and separating the braid, strip and combine first the two black wires and then the two white wires. The resulting pair is a balanced line. This produces common mode rejection at each pin, as well as noise rejection from the balanced line configuration itself. Hum reduction can be as great as 30 dB improvement over standard microphone cable.

Speaker Cable

When audio began to become important, radio manufacturers had a problem. Tubes were used to amplify the signal, and the output impedance of a tube is very high (thousands of ohms). Building a speaker to match that output impedance (for maximum power transfer) would

Figure 5-3

Figure 5-4

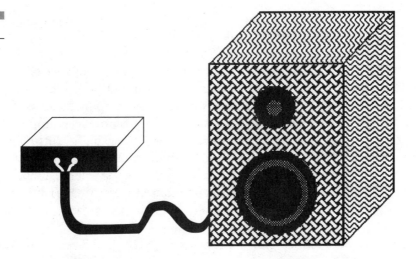

require a large, heavy coil of wire in the speaker, a technically difficult and expensive problem. The solution was to insert a transformer between the amplifier and the speaker, which reduced the high-impedance output to 4, 8, or 16Ω. This meant that only a few turns of wire in the coil of the speaker would match that impedance.

As long as the speaker was inside the radio and the distance from the amplifier portion of the radio to the speaker was just a few inches or feet, it worked well. When hi-fi began to arrive, engineers realized that speaker performance could be improved if the radio (or, more correctly, the amplifier in the radio) and its speaker were in separate boxes. Then the wire to the speaker might run many feet, and the resistance of the wire would be a major factor in using up the amplifier's power (Fig. 5-4).

Because the output impedance of an amplifier is low (4 to 16Ω), the current (amperage) must be high in order to get a given amount of power to the speaker. The lower the resistance of the cable, the more power gets to the speaker (see Appendix A, "Notes and Comments," item 11). It is now fashionable in hi-fi circles to use very large-gage wire to keep the power loss in the wire to a minimum.

The demand for exotic, low-resistance speaker cables has spawned an entire industry. Whether the exotic constructions and materials make a difference, or are worth the extra cost, is still a topic of discussion. But large-gage wire is certainly better than small gage-wire at transferring power from a low-impedance output amplifier to a speaker.

Table 5.3 shows the percentage losses to expect, and distances you can go, at 4 or 8Ω load impedances.

There is a second concern with speaker wiring, called *damping factor.* This is the ratio between the output impedance of the amplifier and the impedance of the speaker. Damping factor can change over frequency. As the frequency gets higher, the damping factor will get smaller.

The actual output impedance of modern amplifiers is very low, and therefore the damping factor is very high. Cutting-edge amplifiers can have a damping factor as high as 20,000 at low frequencies, dropping to 2000 or so at 1 kHz. This allows the energy to be transferred almost entirely to the speaker. Because the vast majority of energy is in the low frequencies, you might think that only the low frequencies are affected, but this is not so.

Much music contains very sharp spikes or fast transitions, such as percussion instruments or piano music. These are, by definition, high frequencies. Indeed, some of these transients and harmonics of music go beyond the range of human hearing, yet many listeners claim they can hear the difference when these ultrasonic signals are missing.

If you could see these sound waves, or their electrical analogs, you would see a very sharp transition from low to high level. This is also

TABLE 5.3

	Distance in Feet at 11% Loss (0.5 dB)						
	24 AWG	22 AWG	20 AWG	18 AWG	16 AWG	14 AWG	12 AWG
4Ω	10	15	25	40	40	90	140
8Ω	20	35	50	85	115	185	285
	Distance in Feet at 21% Loss (1 dB)						
	24 AWG	22 AWG	20 AWG	18 AWG	16 AWG	14 AWG	12 AWG
4Ω	25	35	50	90	125	195	305
8Ω	45	70	105	190	250	395	610
	Distance in Feet at 50% Loss (3 dB)						
	24 AWG	22 AWG	20 AWG	18 AWG	16 AWG	14 AWG	12 AWG
4Ω	85	135	195	340	470	740	1150
8Ω	170	275	390	685	935	1480	2285

called *rise time*. When a waveform with a very fast rise time appears, a poor damping factor will have the effect of smoothing over that spike and therefore affecting the ability of the speaker to reproduce it. The resistance of the wire reduces the damping factor, so the lower the resistance of the wire, the better chance the speaker will have at reproducing such details. Here again, larger wire is better than smaller wire.

You should be aware that very few speakers have linear impedance over the range of audio frequencies. That is, it is rare that they are the same impedance at all frequencies. This is especially true where multiple drivers (woofers, midranges, tweeters) are used to reproduce the entire range of frequencies. Passive circuits are included, whose purpose is to divide the frequencies among the speaker elements. These are called *crossovers*.

Often during the transition between speakers, the apparent impedance can fall from 8Ω to only 2—3Ω. Some amplifiers cannot drive these low impedances without distortion. It takes four times the current to drive a 2Ω load as it does an 8Ω load. In such cases, added resistance in the wire actually can improve speaker performance by providing a greater load.

Distributed Loudspeaker Systems

Impedances of 4 or 8Ω are most common in amplifier outputs. Manufacturers of commercial amplifiers might also supply an output transformer with a number of secondaries, say one marked 8Ω and another marked "70-volt."

The actual impedance of a 70-volt output transformer depends upon the wattage of the amplifier. A 25-watt, 70-volt amplifier has an actual output impedance of 196Ω. There are less common 25- and 100-volt systems. You can go much, much farther with these systems on smaller, cheaper wire than with 4 or 8Ω output impedances. This is because the resistance of the wire becomes only a tiny fraction of the output impedance of the amplifier. Increasing the impedance of the amplifier reduces the amount of current running on the wire. Since this is critical to loss on the wire, increasing the impedance increases the voltage and reduces the loss, so smaller-gage wire can be used (see Appendix A, "Notes and Comments," item 11).

Actual speakers are still 4 or 8Ω, however, so when the wire is connected to each speaker, there must first be a small transformer to reconvert from 70-volt to 4 or 8Ω. Because of this transformation/ retransformation process, the quality of the eventual audio can depend to a great degree on the quality of the transformers. Low-frequency response (below 100 Hz) is lost first. It contains much of the electrical energy, so transformers must be extremely well-made to pass these low frequencies. Because the speakers in distributed systems are most often small, cheap speakers with no baffle or box behind them, the loss of low-frequency reproduction is usually accepted.

Distributed systems are used extensively where quality is not the primary consideration, such as in public address, background music, or paging systems. Dozens, even hundreds, of speakers can be run off one amplifier, and virtually any kind of wire can be used. In background music or paging, the speakers are most often mounted in ceiling tiles or hung on walls. Any kind of speaker could theoretically be wired up to a distributed system. The quality of the original program material (such as background music) is often equally low, however, so putting in high-quality speakers or high-quality transformers will not improve anything.

Another advantage of 70-volt systems is the transformer on the speaker, which comes with various connections (called *taps*) that allow the power fed to that speaker to be controlled. Some speakers can be louder, some medium volume, and some softer, and they will maintain that ratio among them no matter what the setting of the master volume control. This is often desirable where the volume should be high in noisy locations and low in quiet locations.

The main wiring consideration is often the fire rating of the cable. Most speaker systems are in drop ceilings, which normally require plenum-rated cable.

Table 5.4 shows the distance you can run in a 70-volt system at three levels of loss and seven wire gage sizes.

If you want excellent reproduction and cannot accept the compromises of a distributed system, there is one other way to run long lines with speakers. That way is not to run them! Put the amplifiers next to the speaker and run line-level audio down foil-shielded pairs to those amplifiers. They may have to be modified to accept a balanced-line input, but the distance it can run, and the resultant quality, will be excellent. Systems such as this are built into stadiums and auditoriums. Often the remote amplifiers turn on and off with the control console and other equipment at the control point.

TABLE 5.4

Distance (ft) at 11% Loss (0.5 dB)						
24 AWG	22 AWG	20 AWG	18 AWG	16 AWG	14 AWG	12 AWG
520	820	1170	2070	2840	4490	6920

Distance (ft) at 21% Loss (1 dB)						
24 AWG	22 AWG	20 AWG	18 AWG	16 AWG	14 AWG	12 AWG
1120	1770	2520	4450	6100	9650	14,890

Distance (ft) at 50% Loss (3 dB)						
24 AWG	22 AWG	20 AWG	18 AWG	16 AWG	14 AWG	12 AWG
4210	6650	9500	16,720	22,950	36,300	56,000

Digital Audio

The recent move by manufacturers into digital audio has caused a lot of confusion among audio professionals. Some of this confusion is due to the fact that digital audio is partly digital and partly audio. It is digital in that you are converting analog signals into digital data at one of a number of data rates. This conversion produces a high-data-rate signal, where before you had a relatively low-frequency analog signal.

In digital audio, analog rules no longer apply. The signal is now data and must be treated as data. For instance, the cable must have a specific impedance. Cable capacitance is now a critical number. Wire twisting, wiring spacing, dielectric compounds, and shielding style are now more important.

The Audio Engineering Society (AES) and European Broadcast Union (EBU) have set up specifications for digital audio cable. By the time a one- or two-channel analog signal is converted to digital data, there are some significant data rates to be handled. The AES/EBU specifications are:

- 110Ω ±20 percent
- 3.072 Mbps data rate (one or two channels)

The two-channel version is the same cable as the one-channel version. Thus a one-channel cable is also a two-channel cable, and a 12-pair digital audio snake can be a 12- or 24-channel digital audio snake.

The AES/EBU specification mentions an impedance of 110Ω but does not actually give a tolerance. However, ±20% can be extrapolated from the data on equipment within the specifications.

There are three ways to make digital audio cable: immobilize the twisted pairs, use solid pairs, or convert to coax.

Immobilization Take a standard twisted pair cable, use a better dielectric, and then pack the construction with "fillers" (Fig. 5-5). Immobilization keeps the wires at a predictable distance (and therefore a predictable impedance), regardless of the flexing of the cable. Naturally, fillers will add cost and size to the cable. Immobilization also improves the *capacitance unbalance*, which is variations in capacitance over the length of the cable.

Immobilization can also be accomplished by bonding, gluing, or extruding the twisted pair together during the manufacturing process. These constructions have excellent impedance tolerance and low capacitance unbalance.

Solid wires Solid wires can be twisted together with a fair amount of accuracy. They have poor flexibility and flex life, however, and should be considered only for permanent installations.

Digital audio on coax AES/EBU-spec digital audio also can run on coaxial cable through the use of a transformer called a *balun*, which converts from a **bal**anced line (twisted-pair) to an **un**balanced (coax). Because impedance tolerance is so much tighter on coax, you can go up to 500 meters.

Figure 5-5

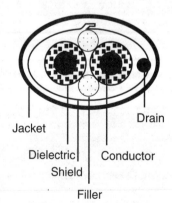

Jacket

Drain

Dielectric Conductor

Shield

Filler

Many of the advantages of twisted pairs are lost, however: balanced lines, common-mode noise rejection, cross-talk rejection, and independent shielding. In addition, two transformers (baluns) must be bought for every channel to be converted, one for each end of the cable! These two transformers easily can cost $40 and up per pair.

If an installation is digital audio and video, it can be cost-effective to put everything on the same coax as the video. Simply buy more coax to do the digital audio, and use the same strippers, crimpers, and connectors. In addition, the labor cost saving of making up a crimp-crimp BNC connector, versus soldering an XLR, might just make the cost of baluns a wash.

SP-DIF and SDIF Digital Audio

SP-DIF and SDIF are two other digital audio standards. SP-DIF (Sony/Phillips Digital Interface) is based on coaxial cables and BNC connectors. SDIF (Sony Digital Interface) is a consumer system based on RCA connectors.

With SDIF, very short lengths (under 1 meter) probably can be run with standard, hi-fi RCA hookup cables. For greater lengths, true 75Ω coaxial cable is recommended. It should be noted that RCA connectors are not any specified impedance, much less 75Ω, but at short distances (with most likely only two connectors in the path), they have a minor effect on the digital signal.

Jitter

Digital signals are sequences of ones and zeros, looking much like the square wave shown in Fig. 5-6.

As cable length increases, the square wave becomes less and less square. Impedance imperfections (structural return loss) reflect some of the signal back to the source, a phenomenon measured as *VSWR*, or *voltage standing wave ratio*.

Capacitance tends to make the edges of each square ragged (see Appendix A, item 13). The point is that, in a system transmitting a digital signal down a cable, the resultant square wave does not look very square, making it difficult to establish exactly where 0—1 and 1—0 transitions are. This is a timing error and is known as *jitter.*

Figure 5-6

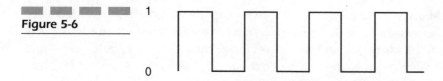

The top graph in Fig. 5-7 shows the original digital signal. The second graph is the same signal after being attenuated by a pure resistance. The third graph shows how the signal might look after traveling down a long piece of wire, or down a short wire of the wrong impedance or high capacitance. The last graph shows the original signal and the distorted signal superimposed so that you can more accurately judge the effects.

Differences in the height of the waveform are attenuation, a reduction in the voltage of the signal. If the resultant waveform were perfectly square, just shorter (lower voltage), that loss would be due only to the resistance in the wire.

R-C Time Constants

Resistance is linear over frequency, that is, it affects all frequencies equally. When you have resistance and capacitance present together, however, you have what is called an *R-C network*. The "capacitor effect" of the cable will charge up and discharge (as seen in the illustration); this effect is made worse by resistance, which lengthens the time that the capacitance charges and discharges. This is why an R-C network is said to have an *R-C time constant*, that is, a charge-discharge time.

There are two ways to reduce the R-C time constant and get our waveform closer to a square: One is to reduce the resistance, and the other is to reduce the capacitance. While it seems like each would be equally effective, reducing the capacitance is, in fact, easier. Here's why.

If we wish to reduce capacitance, we change the plastic on the wire to one with lower capacitance, make a foamed version of the material, or (most commonly) add more of the material we're already using. The wire itself doesn't change. If we wish to reduce resistance, we could change the gage of the wire. While that is simple enough, we also have to maintain the same value of capacitance. With a bigger wire, that means we have to add more plastic to stay at the same capacitance.

In other words, it is easier to reduce capacitance (where only the plastic will be affected) than it is to reduce the resistance, where both the wire and the plastic will have to change.

Attenuation and Jitter

Structural return loss, that is, impedance variations and capacitance, are frequency-dependent. As frequencies (or data rates) get higher and higher, these parameters will affect the signal more and more. Unlike resistance, because they are nonlinear these parameters cannot be ignored. The "non-squareness" of the resultant waveform in Fig. 5-7 and Fig. 5-8 is due to this nonlinear effect.

SRL and other frequency-dependent artifacts show up as jitter, that is, errors in transition between the "up" and the "down" trace. Ideally this transition should be instantaneous. The less perfect it is, the harder it will be for the receiving equipment to determine just how long this bit is. With really bad jitter, the receiving device simply will not function (see Appendix A, "Notes and Comments," item 13).

Figure 5-7

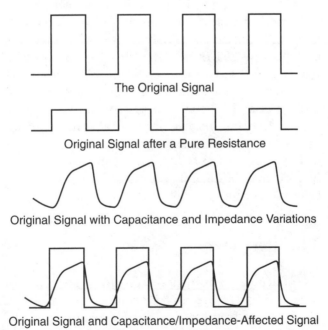

The Original Signal

Original Signal after a Pure Resistance

Original Signal with Capacitance and Impedance Variations

Original Signal and Capacitance/Impedance-Affected Signal

Figure 5-8

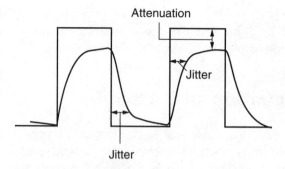

Jitter can be described as a percentage of a signal (e.g., 5 percent jitter), or as a timing error (e.g, 25 ns of jitter). Jitter is measured in nanoseconds (ns) and picoseconds (ps). A nanosecond is one billionth of a second, and a picosecond is 1/1000th of a nanosecond (one trillionth of a second). These may seem like extraordinarily small numbers, but with signals that have lots of information moving very quickly, a small amount of error can have a devastating effect. Since no signal is perfect and every signal has some jitter, the only question is how much jitter a receiver can withstand. The number of nanoseconds difference between the ideal signal and the actual signal describes the severity of the jitter.

Normal digital audio signals have between 300 and 500 ps of jitter. Some hi-fi purists claim they can hear artifacts in digital audio when jitter is below 200 ps. When jitter approaches 40 ns (40,000 ps), most receivers will fail to interpret data with that much error and usually shut down. In the analog world, poor reproduction creates noisy or distorted signal. In digital it just stops working. This can often complicate troubleshooting because it is difficult to determine if there is a bad cable or device or just excessive jitter.

There are two tests that can be performed on digital audio cable to help determine suitability and system performance: the eye test and bit error rate testing.

The Eye Test

To simulate a transmission sequence, a *pseudorandom bit sequence* can be generated by creating an 8-bit computer word that contains all the possible positions from zero to one (Appendix A, item 8). An *eye pattern* (so called because it resembles an eye) is created by looking at these pseudorandom 0—1 transitions on an oscilloscope (Fig. 5-9).

In Fig. 5-9 a *trigger line* synchronizes the scope to the timebase T to keep the eye from drifting across the screen. The perfect eye would show instantaneous rise and fall times between the 0 and 1 states. Impedance mismatches (structural return loss), cable and circuit capacitance, and attenuation all smooth out the sharp edges of the rise and fall, giving the characteristic "eye."

The maximum time difference between the 1 and 0 is called *time jitter.* It is calculated as X divided by T times 100. Most eye pattern tests give numbers in the 5 percent jitter range. Noise margin, shown as H, is the lowest amplitude open eye (or signal peak) measured.

Bit Error Rates

Figure 5-10 is a chart for bit error rates of standard audio cable, impedance-specific low-capacitance AES/EBU digital audio cable, and 75Ω coaxial cable. These are generic results at one-half the clock frequency (3 MHz). Actual results will vary with the quality of the cable, the clock frequency, and the ability of the system to withstand or correct bit errors.

This test looks at the bit errors in a long length of cable by comparing it to a short length. Bit errors are caused by jitter, excessive attenuation, excessive capacitance, impedance mismatch, noise, or virtually any other parameter.

Figure 5-9

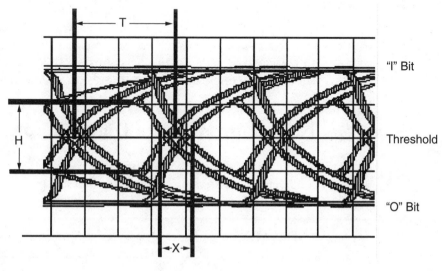

Eye Test Pattern

Figure 5-10

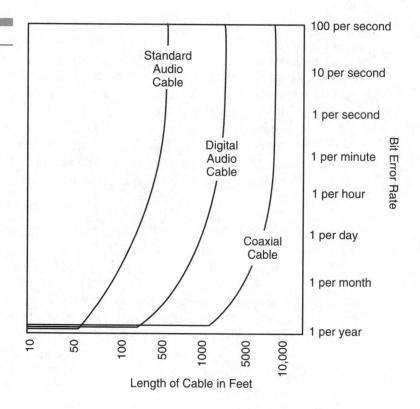

At a given length of cable, the error rate will suddenly start to increase from (for example) one error per day, to one per hour, to one per minute, to one per second. Often, the transition from very few to many bit errors can occur over as short a piece of cable as 50 feet! On the other hand, at a BER (bit error rate) of 10^{-12} (a common BER number), that would be equivalent to one bad bit in almost two days of 6 Mbps digital audio!

New Approaches to Digital Cable

Digital audio designers are looking at computer data cables such as UTP (unshielded twisted pairs) or STP (shielded twisted pairs). These cables have two or more twisted pairs under one jacket and are not individually jacketed or shielded.

The AES/EBU specifications casually mention shielding (or "screening," as is said in Europe). The crosstalk rejection required to isolate one digital audio pair from another is quite low (35 dB minimum). This crosstalk is known in the data world as "NEXT" (*near-end crosstalk*). Where the signals are the strongest (the end nearest the source) is where the crosstalk specifications will be worst.

A NEXT of 35 dB can be met easily by UTP or STP cable known as Category 5. The categories refer to bandwidths; "Cat 5" offers 100 MHz bandwidth. Because we are most concerned with crosstalk at 3 and 6 MHz, a 100 MHz bandwidth is far in excess of AES/EBU data rates.

There is also a new generation of UTP, called *Enhanced Category 5*. These cables offer improved specifications and increased bandwidth. They represent the current limit in standard solid-conductor twisted pairs.

There also are radically different designs, such as Belden's DataTwist® 350 (DT350) with even better NEXT (over 48 dB), extremely low SRL, extremely low cable emission or radiation, very high noise immunity, and very high data rates. DT350 employs bonded pairs, where the conductors in each pair are joined as part of the extrusion process. Impedance variations on such designs look surprisingly like coaxial cable, with average impedance tolerance of ±5Ω.

Having a NEXT of 48 dB when the AES/EBU specifications only require 35 dB is *crosstalk headroom,* an extra margin of safety if the signals themselves are not perfectly balanced.

The Down Side The industry standard impedance of Category 4, 5 or DT350 cables is 100Ω, well within the AES/EBU specification. So why aren't these cables used for digital audio?

- **Unshielded pairs** Because the data pairs are not individually shielded, they cannot be used for analog now and digital later. There is no future-proofing. It's future-only!

- **Balancing** Crosstalk can be caused by twisted pairs in which one conductor is of slightly different length than the other, or in which the twists make variations in impedance, or the balanced signal is not balanced before it gets into the cable. It is possible to electrically balance the signal to make up for some cable imperfections, but it is time-consuming and a needless expense. In DT350, the pairs are bonded together so the impedance tolerance and SRL approaches that of coaxial cable. Signals that enter DT350 in a less-than-balanced state can also radiate. Most manufacturers of AES/EBU equipment are well aware of the critical nature of balanced signals, and most achieve good balanced performance.

- **Solid conductors** Most Category 5 cables (and DT350) are solid conductors. They are installation cables only, and will not take very much flexing before they break.

- **Stranded conductors** Category 5 stranded cables are intended only as short-length patch cables and do not meet the specifications of regular Category 5. Ask for a data sheet and see what the performance is like at 3 MHz and 6 MHz. Crosstalk and SRL are the first two things to suffer.

- **Connectorization** Category 5 cables are intended to be installed much like telephone wires. They are punched down with a special tool into blocks of connecting pins. When the wires are punched, the pins cut through the insulation to make the connection. This is called *insulation displacement*. Category 5 and DT350 are punched down on special high-speed blocks or connectors. Blocks that force the separation of the wires of each pair will increase crosstalk dramatically. In fact, Category 5 pairs should not be separated more than 0.5 inch to maintain full parameters.

The Up Side Category 5 cables exceed AES/EBU requirements for crosstalk, impedance, impedance tolerance, and attenuation. DT350 exceeds them even more.

Connectorization is very fast. Like phone wire, a large number of connections can be made quite rapidly and accurately. This style of connectorization is an outgrowth of the connectors (called an RJ-11) commonly used on all telephones today. The data version is an 8-conductor (4-pair) connector and is called an RJ-45 (Fig. 5-11). This style has been around

Figure 5-11

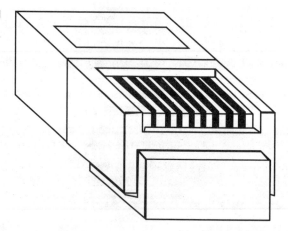

Figure 5-12

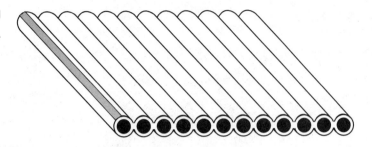

for a long time and has proven its ruggedness and reliability. Cost per pair is very low compared to AES/EBU cable or coax.

As long as the installation is digital-only, Category 5 techniques may improve performance and may provide cost savings.

Flat Cable and Audio

Flat cable, long a mainstay of the computer and data industries, has begun to find its way into analog and digital audio (Fig 5-12). Flat cable comes in a large variety of conductor counts and gage sizes. The key advantage to flat cable is the ability to connect a large number of circuits quickly.

There are a number of variations on flat cable. The first is simply a series of conductors extruded simultaneously. There is a stripe down one side so you know which side is up and which conductor is which. The connectors for this are usually insulation-displacement connectors (IDC). They consist of two pieces which, when squeezed together, push a pin through the insulation of each wire, make the connection, and then lock together (Fig 5-13). Other than cutting a square end, there is no other preparation for the cable. Connecting even a 50-pin IDC connector takes only a few seconds.

IDC connectors are available in a number of styles, the most common being header pins (such as are found on the edge of circuit boards) or subminiature D-shell ("sub-D") connectors, used extensively for computer and control applications. Sub-D connectors are finding their way into more and more audio applications.

This style of flat cable is also available shielded (Fig. 5-14). Shielding usually involves folding an overall foil, under which are one or more loose drain wires. An overall jacket completes the construction and

Figure 5-13

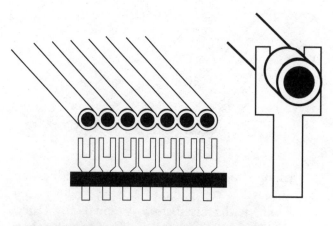

Figure 5-14

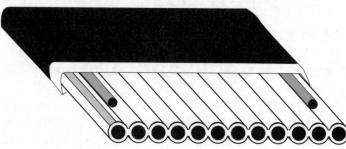

holds the foil in place. This jacket and foil must be cut away to allow connectorization, so shielded versions are more labor-intensive than unshielded in that regard.

There is a second version of standard flat cable, called *rainbow cable* (Fig. 5-15). In this version, each conductor is a different color to aid in identification and connectorization. Because the conductors are different colors, they cannot be extruded simultaneously. Instead, individual conductors are bonded (glued) to a plastic substrate. The color code repeats every ten conductors with the resistor color code (see Appendix A, "Notes and Comments," item 17).

The third version of flat cable has a mesh of wires on one side of the conductors, commonly known as *ground plane flat cable* (Fig. 5-16).

By having a ground element under the conductors, the performance of the individual conductors is dramatically improved, especially in impedance tolerance. The last conductor is offset compared to the others and is in contact with the ground plane; it is, in fact, the ground connection. Although it is offset, it is still near enough to the dimensions of the other conductors that it can be connected with an IDC.

Ground plane flat cable has the added advantage of shielding layers in a stacked configuration. Where there are multiple layers, each successive layer is shielded from the next layer by the ground plane (Fig. 5-17).

Ground plane flat cable is often used in analog and digital console manufacturing for short runs. Ground plane flat cable takes more time to prepare because the ground plane must be separated and cut back before the IDC connector can work. While the ground plane improves impedance stability, it doesn't overcome the main drawback of flat cable: crosstalk.

Crosstalk and Flat Cable

Taking care with how signals are applied to flat cable can improve crosstalk performance. Figure 5-18 shows a standard piece of flat cable. In an unbalanced circuit, with one signal conductor and one ground conductor, the conductors would be designated signal-ground, signal-ground, signal-ground.

Figure 5-15

Figure 5-16

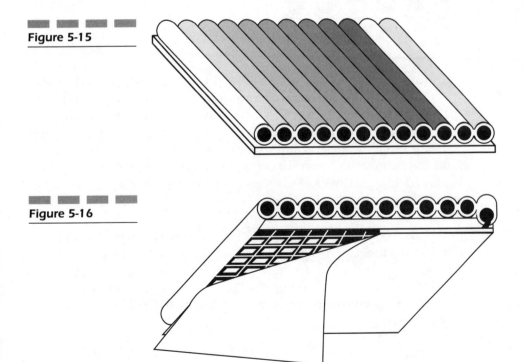

Figure 5-17

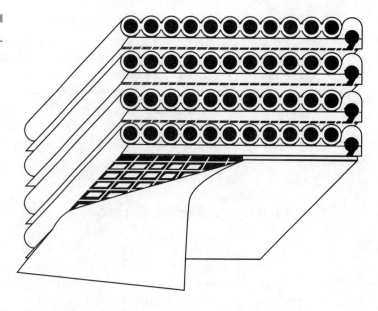

Figure 5-18

S   G   S   G   S   G

In this case S (signal) is always followed by G (ground). Thus the signal wires are physically separated from one other. Separation is always the easiest way to reduce crosstalk. With this kind of flat cable, it is, in fact, the only way to reduce crosstalk. These wires are not twisted together; they are simply straight wires. Crosstalk will always be "poor" with this kind of flat cable and, therefore, its application as a signal-carrying cable is limited. It is suitable for very short runs, or for control signals (on/off, start/stop, etc.). For longer runs, it is limited to low-data-rate digital or low-quality analog.

Unbalanced applications require that you have multiple ground connections to reduce crosstalk. Many circuits would work fine with all signal wires and only a single ground connection, but reducing crosstalk requires that half the conductors be grounds. In other words, your flat cable will be twice as wide as you would really like it to be, a waste of money.

If you wish to wire your standard flat cable in a balanced-line mode, you would want to organize the conductors as shown in Fig. 5-19.

In this version, your balanced lines (+ and —) are separated by grounds. This scheme is somewhat less wasteful than the unbalanced configuration, yet every third conductor still must be a ground in order to reduce crosstalk. It is this dilemma that led to the design of another type of flat cable: twisted-pair flat cable (Fig. 5-20).

Twisted-pair flat cable consists of individual pairs that are twisted and then bonded (glued) to a sheet of plastic. Twisted pairs decrease crosstalk dramatically. To further decrease crosstalk, the pair twists have reversed lays. That is, one pair is twisted clockwise, the next pair counter-clockwise, the next pair clockwise, and so on.

While twisting the pairs reduces crosstalk, it also makes it impossible to connectorize in the same manner as regular flat cable. Since the pairs are twisting, there is no place that an insulation displacement connector can be attached. Therefore, every few inches the twisted pairs go flat. This is enough to cut and to apply IDC connectors without dramatically affecting the crosstalk performance of the cable. The most common dimensions are 18 inches of twist followed by 2 inches of flat.

This cable, even with the flat sections, has crosstalk numbers good enough to be used as a data cable well into the megahertz range, and also can be used for analog audio at reasonable distances. It, too, is available in a shielded version.

Figure 5-19

+ — G + — G

Figure 5-20

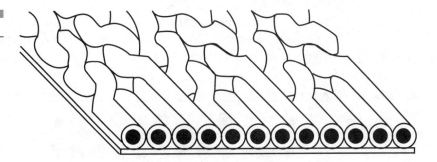

Analog Audio Patch Panels

The piece of equipment that takes more wiring than any other is most often the *patch panel*, sometimes called a *patchbay*. This device consists of rows of connectors, called *jacks*. Sometimes you will hear this device called a *jackfield* for that reason.

Patch panels are very useful devices. When attached to the equipment in an installation, they allow the operator to choose among various devices merely by plugging and unplugging *patch cords* in the patch panel. If a piece of equipment fails, the operator can "patch in" an alternate device or "patch around" the failed device to stay in operation. Patch panels also allow inserting temporary equipment into an installation. All this can be done without opening up any racks, going behind any equipment, or other time-consuming options.

A standard patch panel (Fig. 5-21) consists of two rows of 12 jacks, for a total of 24. Two rows of 13, for a total of 26, is common in video. If more jacks are needed, any number of patch panels can be ganged together. It is not uncommon for large broadcast or recording facilities to have thousands of jacks, or what are sometimes called *patch points*. The panels in patch panels are often 19 inches wide for fitting into a standard rack, with four holes for the mounting screws.

There are a number of variations on patch panels. Some are professional and some are nonprofessional. Table 5.5 is a chart of various connectors available in patch panel format.

The first type of connector, TRS, is the most common patching plug used today. It is virtually identical to plugs and jacks used by telephone operators at the turn of the last century.

There is a miniature version of the TRS plug and jack, called a Bantam (also called TT or "tini-telephone"). These are commonly used in professional installations with large numbers of connections, such as recording studios. Bantam patch panels can fit many more connectors than a TRS panel. In the space for 24 TRS jacks, you can install a

Figure 5-21

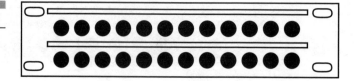

TABLE 5.5

Type of Connector	Normalling?	Balanced?	Professional?
Tip-ring-sleeve (TRS)	Yes	Yes	Yes
Bantam (TT)	Yes	Yes	Yes
Stereo phone	Can be	Can be	No
XLR	No	Yes	Yes
DIN	No	Can be	No
RCA	No	No	No
Mono phone	Can be	No	No

Bantam panel with 96 jacks! Labeling a bantam jackfield is a task in itself, however, because of the limited room. Making up or repairing bantam patch cords requires a fine eye and a good set of miniature tools.

Next is the stereo phone plug, well known as the connector often used on headphones. It also has three connections and is often mistaken for a TRS connector (which it is not). While it can be wired in a balanced mode, and can allow many of the signal routing advantages of a TRS or Bantam, it is not a professional connector and lacks many of the features that give the TRS or Bantam connectors extreme ruggedness and long life.

There are other connectors available in a patch panel configuration (that is, multiple jacks attached to a metal panel). One is the XLR, recognized worldwide as a professional standard connector for microphone and line-level applications.

Another is the DIN plug, recognized as a professional connector in Europe. Versions of the DIN are now commonly recognized worldwide as keyboard and mouse connectors on computers. It is nowhere near as rugged as any of the previous connectors, and often it appears with plastic shells that offer little shielding effectiveness.

Included in the table is the RCA connector, easily recognized as a nonprofessional connector on the back of hi-fi equipment, VCRs, etc. While it is common and easy to find, you will note it satisfies none of the criteria on the chart to be a patch panel connector.

Finally we have the monaural phone plug, a two-conductor version of a stereo phone plug. Because it cannot handle a balanced line, it also has limited applications.

None of these last four is considered a standard for patch panels. The reason is that they do not have the versatility of the TRS or Bantam, or they cannot handle balanced lines, or they lack the ruggedness and long life, or they do not lend themselves to *normalling,* a major advantage in patch panels that is explained later in this chapter.

How Patch Panels Work

A patch panel contains a number of jacks. If it is for professional use, these jacks are balanced-line jacks. Since balanced-line cable consists of a twisted pair and a shield, the balanced jack consists of three connections, two for the twisted pair and one for the shield. The XLR and DIN plugs, though they are balanced-line plugs, are eliminated because they do not allow normalling.

To understand patch panels, you must understand normalling. In the following example, the plugs and jacks could be any of the first three connectors (TRS, Bantam, or stereo phone plug).

Each of these types could carry a balanced-line signal. In each case, the balanced line would be carried by the tip and ring. The sleeve would connect to the shield. The arrangement can be represented as shown in Fig. 5-22.

Notice that the three parts of the plug are labeled the *tip,* the *ring,* and the *sleeve.* This is where the TRS plug gets its name. The jack and plug in Fig. 5-22 could be either the TRS, Bantam, or stereo phone plug described in Table 5.5. The dimensions of the plugs are very different,

Figure 5-22

A Balanced-Line Patch Panel Jack

A Balanced-Line Patch Plug

however, so you must always use the correct jack for the plug you have chosen. The jacks for stereo phone plugs also are much shorter than a standard TRS or Bantam jack. The latter are very long, by comparison, and are sometimes called *longframe jacks* for that reason.

The Bantam jack/plug combination is a miniature version of the original TRS system. While fully professional, the Bantam is very small and hard to wire. Its only advantage is the ability to put many patch points into a small area. Where a TRS patchbay may have 24 jacks on a panel, you can get 48 or even 96 Bantam jacks in the same area. On the other hand, labeling a Bantam patch panel requires labels with type almost unreadably small, or an alphanumeric coding system that requires a book or flowchart just to figure out where to insert the patch cord.

When the plug is inserted in the jack, three connections are made (Fig 5-23). The sleeve (or ground connection) is connected to the tube of the jack. The tip and ring each have a specially designed blade and point that touch the correct place on the plug when it is fully inserted. The tip blade also fits into a slot in the tip, effectively locking the connector in place. This helps to prevent the plug from accidentally slipping out of the jack. Note also the three metal strips, each with a hole, at the back of the jack. This is where the solder connections are made that would lead to the equipment.

If you have a patch panel of all single jacks, such as the one in the illustration, that would be called a *non-normalled* patch panel. With the back of each jack wired to a connector on each piece of equipment, you can establish what is connected to what merely by plugging and unplugging patch cords. This allows great versatility because there is no established wiring pattern. The selection of equipment, and the order in which it is wired, can be changed instantly.

Normalling

One problem with the preceding jack-and-plug setup is that you have to leave plugs inserted to connect anything to anything. This means, even if Box A will be plugged to Box B forever, you must waste a patch cord (or two if it is stereo) to connect them. Your patch panels will be full of patch cords all the time, a messy and confusing situation.

In the ideal world, there is a pattern for equipment. Box A is usually plugged into Box B. It would be nice if you could wire it up that way, inserting a patch cord only when you wanted to change something. That

Figure 5-23

Figure 5-24

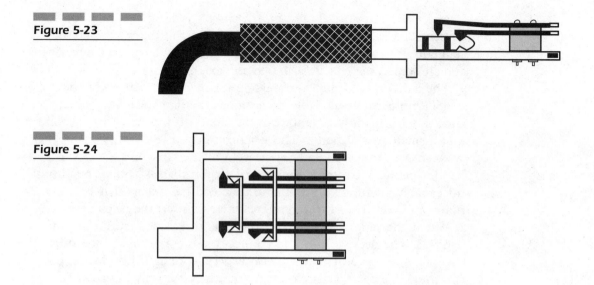

is, if a connection were normal, no patch cord would be required. This can be done by using a *normalled jack* (Fig. 5-24).

A normalled jack actually is a combination of two jacks. They are configured so they will fit exactly into one vertical pair of holes in a patch panel (where two single jacks would have be attached). The normalled jack has two additional switch elements between the tip and ring connections. These blades and points touch and connect the tip of one jack to the tip of the other (and the ring of one to the ring of the other). Thus, signals that enter the top jack are automatically connected to the bottom jack, where they can exit. The only time this connection would be broken is when you insert the plug of a patch cord.

If you insert a patch cord into the top (source) jack, you then have disconnected that source material and can send it to any other destination on the patch panel. If you insert a patch cord into the bottom (destination) jack, you can now choose any other source on the patch panel to be fed to that destination. Since you can interrupt both source and destination, this is called *full normalling*. When only one side is normalled, it is called *half-normalling*. Figure 5-25 shows a half-normalled jack.

In this example, the top jack is switched but the bottom jack is hardwired. This means that if you inserted a patch cord into the bottom jack, the destination would not be broken but would be paralleled. You would have two connections to that destination.

If you turned that jack upside-down, the source would be hardwired and paralleled on a patch cord. Only the destination would be broken by the insertion of a patch cord. This allows you to send the source to more than one device (its usual device and any other of your choosing). Only the destination would be broken.

One other advantage of half-normalling is the ability to use *dummy plugs.* If the source jack is hardwired and you have a patch cord patched into that point, that source feeds some other destination (via the patch cord) and also the normal destination (via the normal connections in the jack below). You can disconnect the normal below by inserting a plug in its jack. If this plug has no wires attached to it, it breaks the connections of the source signal but introduces nothing to that normal destination. Its only purpose is to break that destination, hence the name dummy plug.

When there are normal connections inside the jacks, they are often described as "normals strapped" or "normals strapped at jack." This simply means that the normals have been internally wired.

Panels and Punch Blocks

There are three styles of patch panels readily available, as well as a specialty type of connector called a punch block.

The first type of patch panel consists simply of jacks on a panel, which you can wire up yourself. This is the least expensive but, naturally, the most time-consuming. The second style prewires the jacks and connects them to one of a number of secondary connection schemes. Often these are housed in a box so that the front of the box is the patch panel and the back of the box is the other type of connection.

Figure 5-25

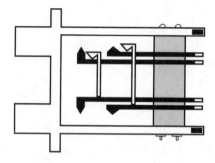

The third style wires the panel to another panel, which contains the secondary wiring part and connects the two with an "umbilical" of wire. This allows you to put the patch panel in one location in a rack and the secondary connection in a different place. Most often, a rack is used that has internal rails facing the back of the rack. The secondary panels are attached to these internal rails.

Punch blocks, often called *66 Blocks*, are insulation-displacement connectors that were originally designed for telephone wiring, which uses solid conductors. Care must be taken when using stranded wire that there are no loose strands to short out or unbalance the circuits.

A special punch tool is required to terminate a wire on a 66 punch block. There are versions of 66 block punch tools that just push the wire into place. These produce unreliable connections. The ideal version is an impact tool which, as you push down, stores the energy and then releases it in one controlled burst, pushing the wire into the insulation displacement terminal. This impact version gives you very consistent, reliable connections.

There is a proprietary insulation-displacement system, the *QCPTM Block*, made by ADC specifically for stranded wire. It does require its own special punch tool, different than other punch tools.

Lastly, there are specialty connections called *Elco/Edac connectors*. These connectors use identical pins (called *hermaphroditic pins*) in both plug and jack. They are very reliable, but expensive. They do allow you to prewire for patch panels and then simply plug the panels into place when they arrive.

Normals Brought Out

The most versatile style of audio patch panel, and the most expensive, is one where the normals are brought out. That is, instead of the internal normal connections being "strapped" inside the connector, they are brought out as separate wires.

If you are wiring up your own patch panel, you can easily bring out the normals if you need to. For prewired patch panels with blocks or connectors, "normals out" or "normals brought out" is an option. While adding to the expense, normals out gives you the option of doing anything. You can easily wire up each jack as a full normal, half-normal, or even run the normal through equipment. If you do wire equipment into the normal, it will always appear between those two jacks—so it had better be what you want!

Self-Terminating Patch Panels

In analog audio, the impedance of the cable is of little importance because the frequencies are so low and the wavelengths so long. However, there is a system impedance chosen as the load required to terminate each signal. The professional analog standard is 600Ω.

When a device is attached to a patch panel, a standard panel provides connections but does not provide the correct load impedance for that circuit. When wired through the normal, to the bottom jack, the destination piece of equipment will provide the load. The source will be correctly matched or *loaded*.

When you insert a patch cord in a jack, however, you automatically disconnect that normal, and therefore you can have no load or, more accurately, "infinite ohms" on that source. When a circuit designed for a certain load impedance sees instead a dramatically different impedance as the load, there is a major level change. If that signal is split from a half-normalled jack, the level on the other side of the split will suffer as well. Great care therefore is taken in audio (and especially in video) to make sure that all circuits are terminated with their correct impedance.

In patch panels, jacks are available that automatically terminate any line. This simply requires a resistance to ground of the correct value, which connects when there is no patch cord inserted (i.e., as part of a normalled switching arrangement).

In a normalled panel, using the correct style of jack will ensure that, when a signal is coming in the back of a source jack, it will be terminated until a patch cord is inserted. The assumption is made that the patch cord will be plugged in somewhere else, where the termination impedance will be supplied by the equipment at the other end of the cord, or will be supplied by the next jack.

Of course, you could still leave a patch cord dangling, and that would constitute an unterminated circuit. A dangling patch cord is pretty easy to see, however, and invites insertion into an appropriate jack.

Project Patch™

For those on a budget, or in a hurry, there is a new type of patch panel arrangement called Project Patch (see Appendix A, "Notes and Comments," item 16). It is a patch panel prewired to accept *header pins*, connectors usually used inside computers and other multiconnection devices. Cables are available in various lengths, with various kinds of connectors on the end. You buy the patch panel and the correct cables pre-made. It all assembles without any tools—except for a screwdriver to attach the patch panel to the rack!

Digital Audio Patch Panels

Digital audio is an area of some controversy when it comes to patch panels. For one thing, we are taking about an impedance-specific system (110Ω), where very low capacitance is required so that the data (essentially square waves) is altered as little as possible.

The Impedance/Capacitance Problem

Standard TRS and Bantam patch panels are not impedance-specific and not low-capacitance. In fact, not one of the patch panel options in Table 5.5 is impedance-specific or low-capacitance. There are three options for solving this problem: ignore it, make something new, or switch to coaxial cable.

The first two possibilities are described in the following sections; switching to coax will be described later.

Ignore the Problem If you have a small TRS or Bantam patch panel, and if you rarely use it for anything but emergency and unusual connections, then the worst-case distance a digital signal would travel in the panel would be in one jack, through the normal, and out another—maybe 6 inches at the most. At digital audio data rates of 3 Mbps, this is not a great distance; a digital signal might withstand such a mismatch, with only a slight penalty in the length of runs to and from that patch panel.

On the other hand, if you already have an installation, it is unlikely that you wired it with AES/EBU 110Ω cable. Therefore there is much more impedance mismatching than just a patch panel. In fact, the patch panel will only make your mismatch worse. While digital audio probably can withstand 50 (or possibly 100) feet of mismatch, it doesn't take many runs back and forth between mixers and processing to get 300 feet of nondigital cable. By that time the signal is so distorted (i.e., the digital jitter is so bad), that the destination device cannot possibly read the bit stream.

Make Something New There are a few proprietary patch panels being made for digital audio. It seems to be impossible to design a true 110Ω patch panel using current patch panel parts. There are some from Europe that have very low capacitance, and others that use an exotic (and proprietary) connector that really is 110Ω—but engineers seem hesitant to lock themselves into a proprietary system. Besides, there are other problems that remain unsolved.

The problem goes deeper than just connectors, impedance, and capacitance. Normalling and half-normalling introduce some daunting challenges to the digital audio patch panel designer.

Digital Normalling

AES/EBU digital audio is really a data signal which sends digital "words" from source to destination. Mixing channels together, as you do in analog, is really mixing multiple bit streams in digital.

Each bit stream contains a *clock frequency,* used to synchronize multiple bit streams. If you are sending multiple bit streams (i.e., multiple channels of digital audio), they may take different paths to get to the destination. If one bit stream goes through a patch panel that is normalled, while another does not (or goes through a patch cord), there will be a difference in distance traveled between those two bit streams. At the other end, the destination device might not be able to synchronize the clocks of the two signals so they can be combined.

If you are half-normalling, things get even more complicated. On the hardwired side of a half-normalled jack, you are adding another path for another signal. What is actually occurring is that you are dividing a signal path; you are creating a **Y**. Two things happen at a **Y**. First the voltage is halved, and second, the impedance is halved.

In the analog world, neither of these is significant. Half of the signal simply means a 3 dB loss in level, easily compensated for. And the impedance of a wire (or patch panel jack or patch cord) is meaningless at analog frequencies. Not so in digital!

If you start with a healthy 5-volt AES/EBU signal, half of that (2.5 volts) is probably recoverable. If you start with much less, say 1 volt, half of that may be difficult to recover, especially if you are going down long lines or through many patch points.

Much more devastating to a digital signal, however, is the division of impedance. First, standard analog jacks have very low impedance, around 40Ω or so. If that weren't bad enough a mismatch for a 110Ω signal, splitting that to 20Ω will certainly make the situation worse. The impedance mismatch will reflect even more of the signal to the source, making the digital signal unreadable through only a few feet of cable.

Digital audio patching also presents a problem in termination. Frequencies are much higher, and therefore impedance in cable is more important, so termination of each circuit into 110Ω is critical for good performance. In a digital audio patch panel, automatic terminating jacks that terminate into 110Ω are essential.

Active Patch Panels

One manufacturer (see Appendix A, item 16) does offer a "regular TRS or Bantam" patch panel for digital audio that features half-normalling. The trick is that the patch panel is active, i.e., it contains powered circuitry to accomplish the splitting and combining of half-normalling.

Such panels also can compensate for timing differences by electrically delaying signals, essentially mimicking electronically the distance the digital signal would have traveled through a patch cord.

Digital Audio and Coax Cable

One solution, and the only one that seems to solve most of the problems, is to convert to coax. If you have dozens or hundreds of signal paths, however, putting a $20 to $40 balun on each input and output can get expensive in a hurry.

The advantages to coax are straightforward. First, you can go farther once you have converted with baluns from 110 to 75Ω. Most precision video cable is low in capacitance, under 20 pF/ft, and impedance tolerance (variations in impedance) is better than twisted pairs, certainly better than the current generation of AES/EBU cable.

Impedance-specific (75Ω) video patch panels are readily available in normalled and non-normalled versions. The only thing you will sacrifice is the common-mode noise rejection that's characteristic of twisted-pair and half-normalling patch panels available in the analog world.

There's an added incentive, too. If you are doing an analog video and digital audio installation, you can use the same cable, connectors, patch panels, and distribution amplifiers for both the audio and video—not to mention the same stripping tools, crimping tools, labels, wire ties, etc.

Splitting Audio

It is a common need in audio to split signals. Sometimes it is to feed a certain signal to two different locations. A certain signal might be required in two control rooms, for example, or an audio signal in a stadium or auditorium might be split to feed both the public address system and a recorder of some sort.

Analog Splits

While a signal can be split by simply creating a **Y** and joining the one input to the two outputs, doing so has two drawbacks. First, anything that happens to output 1 will affect output 2. That is, if there were a short circuit on the first output, the second output also would go dead. The second problem is level. When you split one signal into two, each destination will receive half the intensity (i.e., it will be down 3 dB).

There are three ways that these problems can be solved, with varying degrees of success. The first is a *resistive network*. This allows a constant impedance to each destination. It also prevents a dead short on one side from shorting out the other side completely. However, it does nothing to prevent the 3 dB reduction in signal.

The second solution involves a transformer. This transformer has one primary (input) coil and two secondary (output) coils. Each can be wound to have any impedance required. Shorting one output coil has no effect on the input or the other output coil. Transformer splitters are often used because they solve all the problems except the —3 dB level loss.

The third solution also addresses the situation of having more than two splits. For instance, let's say you have a sports stadium and you wish to feed certain audio signals to anyone who comes to the park: radio stations, television stations, reporters, etc. You might have a dozen feeds of the same material. Splitting a signal a dozen times reduces the level so far that it might be difficult or impossible for each of those users to reamplify the signal.

Therefore the solution is to amplify each split of the signal with a device called a *distribution amplifier*, or *DA*. DAs are more expensive than a resistive network or a transformer. Because a DA is an active device it is prone to failure, where other, nonpowered passive devices give excellent reliability (see Appendix A, "Notes and Comments," item 23). Distribution amplifiers are extremely popular, however, because they solve every problem you might encounter.

DAs can be configured so that every destination receives its signal at the same level as the original (called *unity gain*). Second, they easily can be designed to match impedances in both input and output sections. Finally, the individual outputs can be isolated so that something that happens on one particular output has no effect on any other output, or on the input.

This quality can be especially useful in applications like the stadium example. You have no idea what kind of equipment all these other people will be bringing, what condition it is in, or how well it's been maintained. By providing individual isolated outputs, however, that no longer matters. Whatever effect any one has on its output will have no effect on any other.

Digital Splits

Splitting digital audio is another matter. A regular **Y** cable is out of the question because a split of impedance will dramatically affect the level. A mismatch that great will reflect signals back toward the source and may even render a system inoperative.

Level is less a concern with digital than with analog, but impedance certainly is. Surprisingly, this allows a transformer-based splitter to be constructed for digital audio (Appendix A, item 22). The requirements are fairly predictable: 110Ω input and output, and a minimum bandwidth of 3 MHz. Connectors in and out should be XLR or, for use with coax-based systems, BNCs.

Video

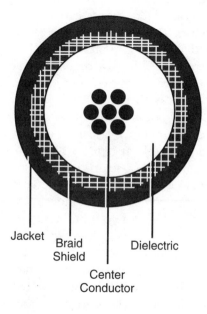

Jacket

Braid
Shield

Center
Conductor

Dielectric

Analog Video

The DC to 4.2 MHz bandwidth of analog video is quite broad compared to analog audio's 20 Hz to 20 kHz.

Historically, coaxial cable has been used for analog video for a number of reasons. A key factor in the superiority of coax cable is impedance tolerance. This is the range of impedance variation that a cable will exhibit over its length. No cable, twisted-pair or coax, is exactly the impedance in its specification for its entire length.

There is, rather, a range plus-or-minus some number of ohms (±) that indicates how close that particular design stays to its intended impedance. Precision video coaxes can be made to a tolerance ±3Ω or even ±1.5Ω. Much of this stability is due to uniformity of the dielectric constant, the diameter consistency of the extruded core, the consistent shield coverage, and the low center-conductor migration.

Hard polyethylene is excellent at preventing center conductor migration, and cables with nominal impedance tolerances of ±1.5Ω are possible. Foamed dielectrics are less effective at preventing movement, and usually are around ±5Ω. Recent work on hard-cell foams has improved their impedance tolerance to ±3Ω error. It should be noted, however, that the low attenuation of foam dielectrics at very high frequencies or data rates far outweighs their loss in impedance tolerance.

Analog coax comes in a number of sizes and styles. Baseband video (as used by analog cameras, processors, switchers, routers, recorders, and players) is a relatively low-frequency band from dc to 4.2 MHz. For baseband coax you must use an all-copper center conductor to obtain the best low-frequency performance, and have a braid or double-braid shield because braid is most effective at low frequencies.

For noncritical, short-distance applications, you can use standard copper-clad steel ("RG-59") coax. Just be aware that the low-frequency performance is compromised and the shield coverage is often less than ideal. If the coax is aluminum wire throughout, it is low-quality cable and should be used only for cable TV (CATV) or similar purposes, never for video.

In general, the larger the coax, the lower the loss and more stable the impedance. Larger coax will probably be more expensive and less flexible.

RGB and Timing

Analog video signals, which consist of black-and-white information, color information, and other signals, are usually sent combined as *com-*

posite video. One way of maintaining the quality of a video signal, especially when it is being manipulated, processed, or recorded, is to separate it into its component parts. This is called *component video* or *RGB,* and it is often used for graphics, animation, or similar tasks where the picture is being electronically created or manipulated.

If you look very closely at the screen of a color television, you will see groups of red, green, and blue dots or *pixels.* RGB stands for the red, green, and blue primary colors on the television screen. There is also a fourth signal, sync. *Sync* is a pulse of the clock frequency (3.58 MHz), which allows the resultant picture to be recombined and included with other video images. All other pictures also contain that standard sync pulse. Without it, every time a new picture or cut is shown, the picture rolls as it looks for the new sync pulse to align to. With one standard sync pulse, you can switch among pictures without any searching for the sync pulse. It's the same sync pulse on everything.

RGB can be sent two ways, by groups of coaxial cables or by individual coaxes. Grouped coaxes are usually called *RGB cable* and come in 3-, 4-, or 5-coax bundles (Fig. 6-1). The coaxes are usually coded red, green, and blue, but otherwise they are identical. The sync pulse is often included with the green signal, or the cable can have a fourth coax (white) and send the sync pulse separately. This is sometimes called *RGBS cable.* A fifth coax (yellow) is also available if a signal is required to run with the RGB. This fifth signal can be one of many things and there is

Figure 6-1

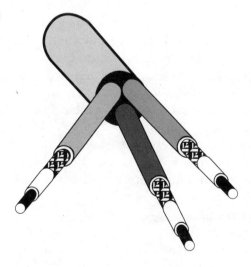

no standard use for it. Sometimes horizontal and vertical sync are sent separately on the third and fourth coax.

Variations in impedance, capacitance, velocity of propagation, and other factors can affect the speed at which a signal can pass down each coax cable. Bundling cables can add *timing errors* to the resultant picture. Timing errors, when they are bad, look like colored ghosts on the screen. Bundled RGB cable can have significant timing error. But if the cable is very short, the timing error may be unnoticeable.

With longer runs, however, check the specifications on the RGB cable. A timing error of 40 ns is considered "not broadcast quality." If there is 1 ns per foot of difference between the multiple coaxes, the signal can only go 40 ft before there is unacceptable quality. Many manufacturers make their RGB "pretimed," that is, the cable introduces very little timing error, or at least too little to see. Pretimed cables are made to have maximum timing errors of between 5 to 10 ns per 100 ft. If a manufacturer says pretimed but doesn't show the specification of nominal timing error, insist on knowing what the error is before deciding which cable to use.

With a cable that has 5 ns per 100 ft of timing difference between any of the coaxes, you could theoretically go 800 ft before reaching the 40-ns quality wall. In truth, you probably couldn't go much beyond 250 ft— and not because of timing, but because of attenuation. Usually the coax in RGB cables is very small. Very small coaxes, by their nature, have higher attenuation than larger cable. A video signal would run out of signal strength through attenuation long before the timing errors became unacceptable.

Use larger individual coaxes in runs over 250 ft, since larger coax has less attenuation and can go farther. Buy these with red, green, and blue jackets to identify which signal is which. Once the cables have been laid and dressed and installed (made neat and tied down), the cables need to be timed. This can be done with a *time-domain reflectometer* (TDR), which shows the electrical length of a cable. The TDR can measure each of the three cables.

Note which cable is electrically "shortest" and cut the remaining two to match that electrical length as indicated by the TDR. (Naturally you should make sure that the shortest reaches the equipment with some to spare!)

Since variations in manufacturing can affect timing, you should try to obtain cables made at approximately the same time. The manufacturer or distributor from whom the cable was purchased should be able to identify cable made at the same time. This will help reduce variations in parameters. If the cables are seriously different in manufacturing, or if the runs are thousands of feet long, you might easily have one of the three legs several feet longer (electrically) than the other two. This can make a neat installation difficult, because it is essential that that extra length be maintained regardless of its appearance.

Where a TDR is not available, a *vectorscope* can work almost as well. This device shows the relationship between red, green, and blue signals. It does not indicate length, however, just timing. A color bar signal must be introduced in one end of the cables to give the vectorscope something to measure at the other end. You then can determine easily which run is electrically longer or shorter, but the vectorscope will not tell how much to cut off to line them up. This means that a piece at a time will have to be snipped off and the shortened cable reattached to the vectorscope after each cut. One trick is to use screw-on BNC connectors, which can be easily screwed on after each cut. (Screw-on BNCs are not recommended for permanent installation, however, the crimp-crimp connectors are much more reliable.)

Analog Video Patch Panels

Patch panels are commonly used in video to route and control signal flow throughout an installation. Even where expensive routing switchers are used, critical functions are often backed up by patch panels. Patch panels are passive. They do not stop working when there's a power failure. They are proven technology—the audio version dating to before the beginning of this century!

Video jacks are usually arranged in pairs, although single jacks are available. The connector at the end of each jack is usually a BNC, with the shell of the jack connecting to the shell of the BNC (Fig. 6-2). Usually

Figure 6-2

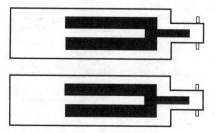

the ground or shell is common to every jack. That is, they are all mounted on a metal frame and therefore are all connected. Only the center pin remains unconnected.

To make a connection in this arrangement, a patch cord would be inserted in the two jacks, completing the connection. The patch cord has a special 75Ω coax connector, designed for thousands of insertions and disconnections (Fig. 6-3). To complete a circuit, therefore, a patch cord would be inserted between two jacks. This would connect the cables attached to the back of the jacks (Fig. 6-4).

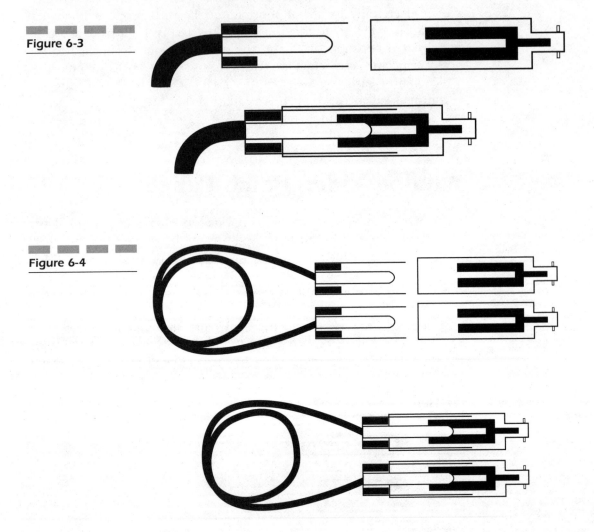

Figure 6-3

Figure 6-4

Figure 6-5

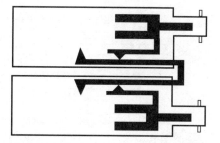

Usually the top row is sources, that is, where signals are coming from. The bottom row is usually destinations, where signals are going. There is no industry standard, however. Often there are many more sources than destinations (and occasionally vice versa), so you might see a panel with all single jacks that are all sources. For instance, an installation such as a stadium, auditorium, or fairgrounds might have multiple tie lines from the outside world, or from satellite downlinks. These would most often be source-only, with no return circuit.

Normalled Patch Panels

Where patch panels consist of single jacks or dual jacks that are not electrically connected, they make up what is called a *non-normalled patch panel,* that is, there are no default connections made by the panel. There are special jacks called *normalled jacks,* however (Fig. 6-5). In these jacks, there is a switch in each connector. This switch attaches the center pin in one jack to the center pin in the jack above or below it.

Note that the center pins of normalled jacks are connected together only as long as a patch cord is not inserted in either jack. As soon as a patch cord is inserted, the shell of the patch cord pushes away the switch, which breaks the circuit. This arrangement, where both jacks are normalled and a patch cord into either one can break the circuit, is called *full normal* (Fig. 6-6).

Half-normalled jacks, common in the audio world (see "Audio Patch Panels" in Chap. 5) are not used in video. The reason is that adding a cord to the non-normalled side would radically alter the characteristic impedance of the circuit. Essentially you'd have two cords attached to one jack. That means the impedance feeding each would be 37.5Ω (half of 75Ω). Such an impedance mismatch would cause radical signal loss due to increased SRL, and therefore is not possible.

Figure 6-6

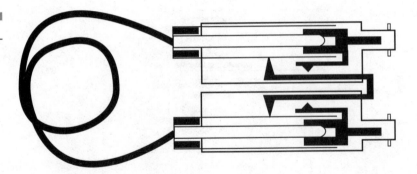

Timed Patch Panels

Things get a bit more complicated when the signals passing through the patch panel are timed signals. Whether they are RGB signals or other clocked signals intended to stay in time with each other, the arrangement shown thus far will not work.

The primary problem is that the distance between the two pins in a normalled jack is a matter of inches. When you plug a patch cord in, the distance between the pins grows to become the length of the patch cord. If you are patching RGB, you have to patch three signals. The timing will be off by the difference of the length of the patch cords.

In composite video, where all picture elements are contained in the one signal, this presents no problem. But in component video or other timed installations, inserting a short patch cord can put that RGB signal outside of the 40 ns timing window for broadcast quality.

The solution to the problem is to give each jack an output for each normal (Fig. 6-7). Then, between these normal jacks, put a coil of cable equivalent to the length of a patch cord (Fig. 6-8). This means that, when you interrupt the signal with a patch cord, a timed signal will not be able to tell the difference because the normal it just passed though is the same length! The key here is that the length of the normal cable is identical to the length of the patch cord (Fig. 6-9).

There are two important steps before ordering and installing these patch panels. First, you must figure out how many of your patch panels need to have this timing feature. Then you must figure out how long the patch cords will have to be to reach from the jacks that are farthest away from one another. In other words, you must figure out the maximum length of a patch cord for the timed portion of your patch panel installation. That length must be the length of the coil of cable between normals. The most common lengths are 3 ft and 6 ft.

While you could buy jacks with timing connections and make your own timing cords and patch cords, most installers buy patch panels that are premade. In the case of timed normal video patch panels, if you want something other than standard-length patch cords (and normals), you must order it specially from the manufacturer.

It is also suggested that the patch cords for use in a timed installation be color-coded red, green, blue, and a fourth color (such as white or yellow). It is also suggested that the labels on the patch panel itself be similarly color-coded. That gives the operator a fighting chance of putting

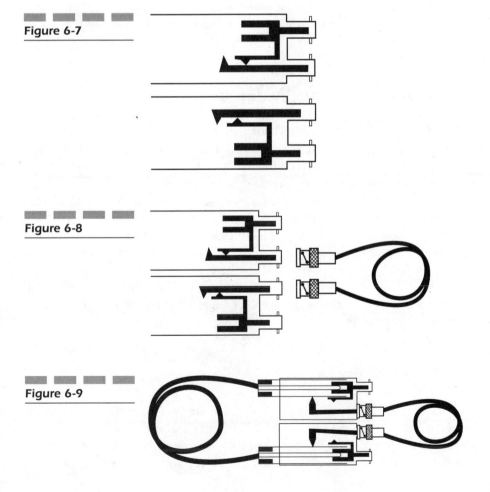

Figure 6-7

Figure 6-8

Figure 6-9

the correct patch cord into the correct jack. Nothing looks stranger than a video picture with switched RGB components!

RGB Patch Panels

Finally, there are true RGB patch panels. These are very expensive patch panels in which the connectors are grouped in threes, as shown in Fig. 6-10. (Some are even available in fours.) They use special patch cords that contain multiple connections. Often the patch cords and corresponding jacks are arranged so that you cannot plug them in the wrong way.

If you have a lot of RGB equipment, that is, if you're doing a lot of graphics, animation, or image manipulation, a true RGB patchbay with the safety connectors might be worth the extra cost, just for the added peace of mind.

Self-Terminating Panel Jacks

Self-terminating jacks, a good idea in audio, are even more important for analog video. If you're installing a complex patch panel arrangement where you cannot predict what will be connected into what, a self-terminating style of jack should be employed.

Self-terminating jacks have a resistor from each pin to ground as part of the wiring of the normal. Anything wired into the patch panel is automatically terminated into 75Ω. When you plug a patch cord into the jack, the resistor will be disconnected. The termination is only broken when a patch cord is inserted. Thus the chances of creating an unterminated line are greatly reduced. For the minor increase in cost, you will not have to worry about dramatic changes in video level or other artifacts of unterminated circuits.

A Patch Panel Caveat

To make matters more complicated, there are two different standards for video patch jacks and plugs. The variation is in the size of the pin in the plug and pin receptacle in the jack. One version is 0.090 inch and the other is 0.070 inch (see Appendix A, "Notes and Comments," item 21).

Figure 6-10

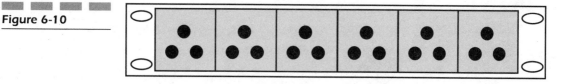

These two are not compatible. In fact, plugging a patch cord with an 0.090 pin into a jack meant for 0.070 will bend—and possibly break—the pin receptor. At the very least, that jack will be intermittent and unreliable for the rest of its life.

The 0.090 size seems to be winning the war of the patch panels, at least in new installations. If your installation is wired with the 0.070 pins, however, you will have to maintain that size throughout. There is no inherent performance advantage of one size over the other, although an edge in ruggedness might be cited for the 0.090 size.

For that reason, be careful when clients or customers bring in their own equipment. Be sure the patch cords you use are yours and not theirs. If you make up some adapter cables from one pin size to the other, keep them under lock and key. Murphy's Law says that if you have 0.070-to-0.090 adapter cords, no matter how well you label them or what garish cable color you use to construct them, someone will take them when you least suspect it and shove them into your patch panel. Lock them away!

Choosing Patch Panel Cable

If you're wiring up one or two video patch panels, choose the best cable for the job and go to it! If you have racks full of patch panels, you will have to do a lot of preplanning or you will have total chaos as you try to wire every jack with one of a thousand cables. There is one important consideration that needs to be addressed now: cable size.

If you wire patch panels with analog precision video cable (such as Belden 8281), you will soon find that the cable size (0.305 inch) is almost as big as a BNC. By the time you have a few rows partially filled, there is no way you can fit your fingers in to attach the last bunch of cables and connectors to the jacks that are left. There are two solutions.

Use a BNC insertion tool. This is a tool that holds the BNC and allows you to insert it and connect it in a space that is only slightly wider than

Figure 6-11

the BNC itself (Fig. 6-11). They come in 6-inch and 12-inch sizes. It's probably a good idea to have one of each size; both can come in handy in a large and complex installation.

Use different cable. Before you resign yourself to using less-than-top-of-the-line RG-59, consider other video cables that are available. Some are even precision video and therefore may be appropriate to wire not only the patch panel portion, but the entire job as well. Some are RG-59 size (such as Belden 1505A) and some are RG-6 size (such as Belden 1694A). Naturally, moving to different cables means different stripping tools, different crimping tools, different connectors, and even different equalization (EQ) cards if long analog runs are being planned. Changing to smaller cables can require major re-engineering to an installation design.

Twisted Pairs

In audio, twisted pairs offer balanced lines, crosstalk rejection, and shielding independent of signal. Only the most exotic and expensive twisted pairs can support analog video bandwidths.

On the other hand, there are a few companies who are producing passive and active boxes to take advantage of the installed base of high data rate twisted-pair cables. These devices attempt to make up for unbalanced parameters, impedance variations, and other twisted-pair phenomena. The distance that analog video can be transmitted by these devices is due, in large measure, to the quality of the cable.

For instance, with an active balanced transmitter and receiver, manufacturers claim to send "VHS quality" analog video down standard phone-style Category 2 pairs for 1000 ft; Category 5 cable, 2000 ft; and DataTwist 350, up to 3000 ft. There are other companies who claim their active balancing is so good they can double those distances. There is more on these cables in the next chapter, "Multimedia."

Triax

Triax cable is used for wiring cameras and associated equipment. Triax starts as fairly standard small (RG-59) or large (RG-11) video cable. Then another layer of insulating plastic is applied, another braid shield is added, and an overall jacket is attached around the outside (Fig. 6-12).

Because of this second braid, even small triax cable can get very large, and large cable can get huge. Triax that's well over half an inch is quite common and, in fact, the size can have some advantages. The main advantage is in the studio, where it connects the camera to some kind of wall plate or junction box.

Cameras are often mounted on movable pedestals, or *dollies*, that wheel around the studio. Around the pedestal is a skirt whose job it is to push the cable out of the way when the camera is moving. Small cable can get under the skirt. This could be disastrous; a heavy camera could destroy triax by rolling over it. If that camera were on the air, rolling over a piece of triax would create a very bumpy shot—rather like driving over a speed bump.

While large size can ensure that the cable is pushed away by the skirt, it needs to be flexible, rugged, and quiet (no squeaking on the floor). These are generally considered conflicting requirements, so it is a major accomplishment that large, flexible, rugged, and quiet triax does exist.

Triax can be very expensive, but it is worth every penny. The reason is that every signal coming from and going to the camera can travel down the triax by a method called *multiplexing*. First is the power for the camera;

Figure 6-12

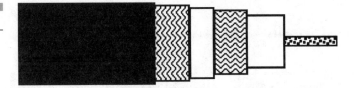

triax often has between 160 and 600 volts on the two braids. Naturally, the video signal is going down the center conductor and the first, or inner, braid. (Yes, the power in and the signal out share the same braid.)

That's not all. The video teleprompter feed, the tally light (which tells the talent which camera is on the air), the cameraman's headset and microphone audio, and even some automatic functions such as lens adjustments, can all be multiplexed on this cable. The cable it replaced had as many as 81 conductors, so the added cost of triax is a small price to pay in streamlining and simplifying camera operation.

S-Video

S-video is based on home videocassette recorders, called *S-VHS* for "Super Video Home System." The main difference between VHS and S-VHS is that in S-VHS the video signal is split into two components to increase quality. The two parts are the *luminance* (the black-and-white information, also called *Y*), and the *chrominance* (color information, also called *C*). The S-video signal is also called *Y-C.*

Because it supports a dual signal, S-video cable is two coaxes. Most S-video cable is low-quality, serve-shield miniature video cable, however— certainly not the quality of precision video cable we have discussed before. The format, and the machines that use it, are not truly professional, and S-video cables are never run more than a few dozen feet. The quality gained from substituting precision video cables might be minor.

Digital Video

There are two systems for digital video, parallel and serial. The system used with a particular piece of equipment is already chosen by the manufacturer. It makes sense to know the pitfalls of the wiring of each system before purchase. Table 6.1 compares the two on several critical parameters.

Parallel Digital Video

Parallel uses 10 pairs, each carrying $\frac{1}{10}$ of the total 270 Mbit signal. The eleventh pair carries the "clock" signal, which allows the various parts of

TABLE 6.1

System	Data Rate (Mbps)	Cable	Distance	Cost
Parallel	27	11 pair	30 m	high
Serial	143—360	1 coax	variable	medium

the video image to be locked back together. While this allows the use of fairly standard cable, connectorization is very time-consuming (22 wires plus a drain wire for the shield in a 25-pin DB-25 subminiature connector). The multipair cables also can be very large (more than 0.4 inch outside diameter).

The variation in impedance (impedance tolerance) is not as good for these twisted-pair cables as it is for coaxial cables. Impedance reflections (SRL) become too great even at the low (27 Mbps) data rate, and cables are limited to a maximum distance of 30 m (98 ft).

Serial Digital Video

Serial digital video, of which there will be much more said in the coming pages, uses coaxial cable. Impedance tolerance of coax is excellent, some cables having impedance variations as low as ±1.5Ω. This is probably as low as any type of cable that's made for transporting video. Capacitance is very low, usually under 20 pF/ft, so you can go much farther on coax than on any other medium.

Connectors used for video on coax are most often BNCs (Fig. 6-13). The BNC is reasonably priced, easy to connect, impedance-specific, and can be made in high-performance versions for digital video.

There is some controversy regarding the choice of connectors for serial digital cables. In the analog world, almost all BNC connectors are 50Ω. The reason is simple: BNCs started out as 50Ω connectors. They were readily available, the tooling was already in place, and prices were low.

As shown in Fig. 6-13, BNC connectors consist of a pin that is crimped or soldered onto the center conductor of the cable. The pin is inserted in the connector shell, and a collar is crimped over the entire cable to hold it in place. In the cutaway view of a 50-Ω BNC (Fig. 6-14), the dotted area is the dielectric, usually polyethylene, which gives the connector its characteristic impedance. The rest is metal. The slot in the dielectric is where the pin is inserted.

Figure 6-13

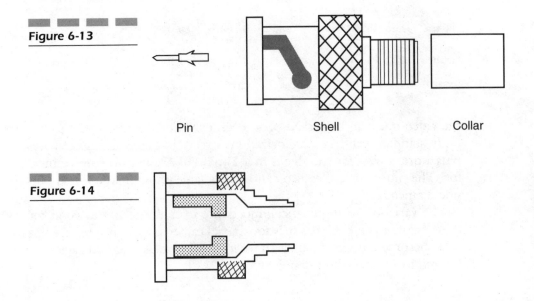

Pin   Shell   Collar

Figure 6-14

Since a BNC has about half an inch of internal working length, a 75Ω cable with a BNC connector at each end has about an inch of total length that is 50Ω instead of 75Ω. At 4.2 MHz (the highest frequency for analog video), that translates to a wavelength of 234 ft. One inch is unnoticeable. Going through patch panels, routers, distribution amplifiers, or processing equipment—a total of (say) 24 connectors is equivalent to about a foot of 50Ω line on a 75Ω cable. Even that amount of mismatch is minor in analog video. The difference cannot even be seen with a TDR.

In serial digital, you can approach mismatch disaster much sooner, if only for the fact that the bandwidth is much higher in digital than in analog. Component digital at 270 Mbps has a bandwidth of 135 MHz, which had a wavelength of 2.22 m (7 ft 3 in). One or two 50Ω connectors in a digital 75Ω circuit might have little effect. Two dozen 50Ω connectors, equivalent to a 1-ft "piece" of 50Ω cable, can be deadly. The equipment at the end of that cable may be unable to receive sufficient signal strength because of SRL reflections.

Analog mismatch shows up as increased attenuation, increased noise, screen ghosts (the reflected and delayed signal arriving after the initial signal), or some other apparent screen anomaly. Digital, on the other hand, is pretty much a go/no-go situation. Only so much correction can be applied before something says, "Sorry, not enough level" or "Sorry, too many bit errors" and shuts down. The first indication of what was a minor impedance mismatch in analog, when translated to digital, might be that nothing works!

75Ω **Connectors**

The solution is simple. Use "compatible" 75Ω BNC connectors. These are BNCs in which the impedance has been modified from 50 to 75Ω by adjusting the dielectric material between the center pin and the shell. In some current connector designs there is no dielectric material ahead of the pin seating slot, as shown on the left side of Fig. 6-15. In others, there is only a thin tube of material, as on the right. It is easy to tell a 50Ω BNC from a 75Ω compatible BNC. Just look at it front-on. If there is a thick white tube of plastic, it's probably 50Ω.

The key to compatibility is that a compatible 75Ω BNC will mate with an old 50Ω BNC. The shell and pin are still the same dimensions. So change over now, use them for analog now, and be wired and ready when that machine, switcher, router, or patch panel is traded out for a digital version.

There is a last word of caution about 75Ω BNCs. In the past, 75Ω BNCs were made for laboratory and other precision uses. In those designs, 75Ω was achieved not by adjusting the dielectric thickness, but by making the center pin much smaller. Such connectors are not compatible with anything but their own kind; a true 75Ω plug will flop around inside a 50Ω jack or a compatible 75Ω jack. If a 50Ω BNC or compatible 75Ω BNC plug is inserted into an old-style 75Ω BNC jack, the jack can easily be broken; it takes a very small pin, while a 50Ω BNC or compatible 75Ω BNC plug has a much larger pin. Be sure to buy 75Ω compatible BNCs. If they don't mate easily with 50Ω BNCs, they're not compatible.

Figure 6-15

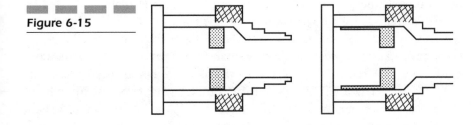

Serial Digital Cable Construction

Since the digital signal extends from zero bits to hundreds of megabits or beyond, a 100-percent copper center (no copper-clad steel) is required. Copper-clad is good only above 50 MHz because of skin effect. In digital video there is a significant percentage of signal below 50 MHz, so use only solid copper.

Stick with a solid center rather than stranded. Solid copper has much more predictable surface area than stranded. A solid conductor therefore has more consistent skin effect at high frequencies. A stranded conductor will eventually exhibit some wavelength-dependent bumps and dips that will become more prominent at higher frequencies. A stranded conductor is more likely to move, spread, or be otherwise unpredictable in its actual position within the cable, especially when the cable is bent. This means that impedance variations will be greater in stranded than solid, which can be a serious problem at digital data rates.

Digital coax also must have an effective broadband shield. Even double-braid shields are 50 percent or less effective at the upper end of the digital band. The ideal shield would be foil-braid, where the braid effectiveness is as close to the maximum (95 percent) as possible. Foil is very effective at high frequencies. The crossover between foil and braid (at about 10 MHz) gives the ultimate in low- and high-frequency broadband shielding.

The dielectric choice is a problem. Use standard polyethylene and it will not have good high-frequency response. It can be used, but the cable will not be able to run very far (generally under 1000 ft). Choose foam, for excellent high-frequency performance, and it may be hard to extrude a core with a consistent dielectric constant. Center conductor migration can be a problem because the foam might be soft. Make the foam a hard-celled foam and the cable will be inflexible. Make the cable big so there's more foam, making the tolerances easier to meet, and it will have a very large outer diameter. All four variations are currently available.

While solid-dielectric video cables can be used for digital, one problem emerges. Foam cables, especially gas-injected foam cables with high velocity of propagation, have much better high-frequency response than standard solid-dielectric video cables. If both cables are being used for analog video, their equalization curves will be completely different. While virtually all manufacturers make equalizers for gas-injected foam cables, matching old and new equalizers with old and new cable could be a nightmare.

A rule of thumb in such cases is this: If it is new construction, or a separate, autonomous standalone portion of a facility, use the gas-injected foam designs. This will allow future-proofing the installation; the gas-injected designs will work well for analog now and will be excellent for digital later. When modifying or adding to an existing facility, cable choice will depend upon how the installation is configured. If there are long equalized lines, it is best to stick with the cable already in use, since any equalizers used are set for that cable type. If it is a very small installation with no equalization, any cable type or combination of types, old and new, can be used.

If the facility is considering digital as a future possibility, using even regular precision video cable now (with compatible 75Ω connectors) could save money later. If you continue to use 50Ω connectors, those connectors will have to be changed when the digital equipment arrives.

Serial digital video differs from analog video or digital audio. Digital audio operates at a maximum of 3 Mbps; standard digital video operates at any number of different data rates. In fact, the proliferation of data rate schemes makes it very difficult for the uninitiated to choose the right system.

In common analog-to-digital (A/D) conversion, the analog signal is chopped up into discrete bits. The frequency with which the analog signal is chopped up is called the *clock frequency*. Since it is the rate at which conversion occurs, the higher the clock frequency, the more accurate to the original the copy will be.

There are compression schemes for digital audio and video that attempt to reduce the number of bits generated, or to reduce the data rate by selection of bits. This allows the data to be recorded, manipulated, transmitted, received, and reconverted with much less channel space or bandwidth. It is still a fact that the lower the compression rate, and higher data rate, the higher the quality. The reverse is also true: A higher data rate usually indicates a lower compression rate.

There are digitizing processes, or *protocols*, for digital video, with data rates as low as **56** kbps and even lower (see Appendix A, "Notes and Comments," item **29**). But only high-megabit data rates will approach broadcast quality.

Transmission rates, that is, the data rate delivered to the consumer, is often much lower than original program rates. For example, the DirecTV satellite sends a 6-Mbps data stream for each channel. While the quality is described as "VHS," and the onscreen noise is greatly reduced because it is a digital signal, it is in no way professional video quality.

Table 6.2 shows the data rates for American (NTSC) and European (PAL) video systems, together with the proposed High Definition Television (HDTV). These data rates are at the cutting edge of today's cable technologies. In the computer world, data rates of 100 Mbps are considered the cutting edge! Digital video is many times that rate. In HDTV, the actual data rate is 1.485 Gbps. Because of that fact, the data rate chosen (360 Mbps) is actually a compression of that data. Sony's proposed 1.485-Gbps uncompressed digital video system cannot run farther than 100 meters (328) on even the best digital video cables currently available.

Bit Error Rate and Headroom

The use of bit error rates (BER) in determining the cable length seems to be the most accurate measurement to indicate how far a digital video signal can go on any medium. In the curve that describes the error rate, there is a "knee" where the BER climbs from "seldom" to "often." This transition can take place in as little as 50 ft of cable. At 800 ft, for instance, there may be unreadably low errors, but at 850 ft you could have one error per frame, a wholly unacceptable rate.

It is difficult to predict BER, even for a single cable design, because of the variables in generating and receiving digital signals, error correction, level sensitivity, and other factors. There is also the question of what an acceptable bit error rate is (Table 6.3). For telcos with T-1 lines, a BER of 10^{-7} is considered acceptable. For digital audio and most data systems (such as ATM), a BER of 10^{-10} is considered acceptable. For digital video and other very high-quality transmission systems, 10^{-12} is usually cited the standard. You will note that the amount of time between errors at 10^{-12} is so long that this is generally considered "no errors."

TABLE 6.2

System	Data Rate (Mbps)
Composite NTSC	143
Composite PAL	177
Component NTSC	270
Component PAL	270
HDTV	360

TABLE 6.3

Bit Error Rate	Equal to one error in	At 6 Mbps (AES/EBU) one error in	At 270 Mbps digital video, one error in
10^{-7}	10 million bits	1.67 s	1.11 frames (30 errors per second)
10^{-10}	10 billion bits	27.8 min	1111 frames (37 s)
10^{-12}	1 trillion bits	46.3 hours	111,111 frames (1 hr 1 min 44 s)

Headroom is a term usually applied to analog signals. *Headroom* is the difference between the level with acceptable signal-to-noise and the level at which recording tape saturates (maximum recording strength), or other similar maximum limitations of transmission, recording, and playback equipment. Headroom is more critical for digital systems. Exceeding the maximum input level will create artifacts that will greatly increase bit error rates, often to the point of system failure.

As a rule of thumb with 270 Mbps digital video, most system designers will allow 6 dB of headroom. They insert a precision wideband attenuator at the end of each video line and adjust the attenuator to add 6 dB of loss to the line while watching the results on a video monitor. Bit errors appear as speckles or flashes on a video monitor. If bit errors appear on the monitor being watched before that extra 6 dB is added, then that cable is too long and the system design is readjusted to make it shorter.

The Digital Video Bell Curve

In digital video, the *occupied bandwidth* or *spectral distribution,* as viewed on a spectrum analyzer, is no longer a series of carriers and subcarriers as it appears in analog video. Instead, it is a *bell curve* of energy (Fig. 6-16) with the top of the bell (most of the energy) at the center of the bandwidth. The point F is the upper limit of information, the highest data rate, or clock frequency.

Anti-Alias Filtering

Some purists, in both audio and video, are upset that severe filtering takes place prior to processing into a digital signal. They argue that in

Figure 6-16

Digital Video Spectrum

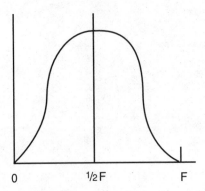

0   ½F   F

analog there is no built-in limit to the bandwidth of a signal (that is, it is limited only by the equipment and personal standards that had been chosen). In digital, however, there is a severe cutoff in frequencies, above which there is no signal at all. When we have cable capable of (almost) unlimited bandwidth, why must this be?

This is because of one artifact of digital processing: *aliasing*. Aliasing requires a filter to prevent it, called a *Nyquist filter*. Digital video requires a Nyquist filter, as do all digital systems. This limits the processed data to frequencies below the clock frequency. If data above the clock frequency were present, the resulting sampling would produce nonexistent waveforms or aliases. This is why such filtering is often called *anti-aliasing*.

Let us assume that the waveform shown in Fig. 6-17 is an incoming signal. We wish to digitize that signal. Therefore we pick a clock frequency by which we will divide up the original wave. Figure 6-18 shows the sample (or clock) frequency. When the waveform is sampled by the sample frequency, the resulting pattern is as shown in Fig. 6-19.

Showing just the four resulting points (Fig. 6-20) makes it look as if the waveform would be difficult to recreate, that there is not enough information to bring back the original. Surprising as it may seem, however, the four points of data are enough to recreate the original waveform with acceptable accuracy (Fig. 6-21).

In fact, many systems sample the highest frequencies at two or even fewer samples per waveform, yet they achieve surprisingly accurate results and low distortion. Let's assume, however, that we introduce a waveform of greater frequency than the sample clock frequency, as shown in Fig. 6-22. Again we sample it with the sample clock (Fig. 6-23).

Figure 6-17

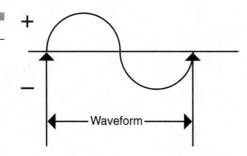

+

−

◀——— Waveform ———▶

Figure 6-18

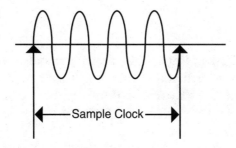

◀——— Sample Clock ———▶

Figure 6-19

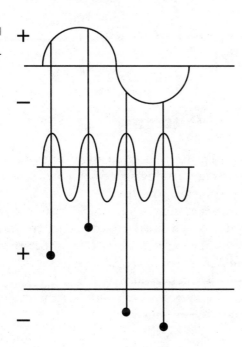

Figure 6-20

Figure 6-21

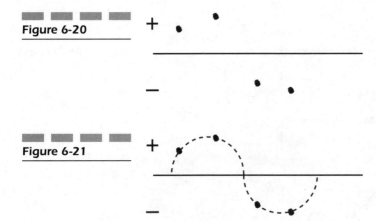

Unlike our first example, this sampled signal, while taken from the original waveform, results in four points that are nothing like the original (Fig. 6-24). It lacks sufficient points to be recreated as an accurate waveform. In fact, when converted back it's a completely different waveform (Fig. 6-25). It's an *alias*.

The alias is always lower than the clock frequency, because the clock frequency cannot sample anything faster than itself. This is why digital systems always have hard, "brick wall" filters that pass nothing above a given frequency. These hard filters are there to prevent aliasing. That is why sampling up to the clock frequency is often called the *Nyquist limit*. To sample those higher frequencies would require an even faster clock.

Video Moiré Patterns

In digital video, aliasing is an even more disturbing problem because you are mixing complex signals together. Often these will interact, producing visual interference patterns.

You may have seen such interference in analog video, for example where a striped tie or a herringbone jacket produces odd, motion-sensitive patterns. These effects are even more prominent in digital video. These interference or *moiré* patterns are avoided easily, however, by setting the Nyquist filter at one-half the clock rate. This prevents any frequencies that are even remotely close to the clock frequencies from getting through to the sampling circuitry.

Digital Video Cable and Distance

Table 6.4 shows examples of three Belden precision digital video cables, together with the effective lengths that can be run at various data rates. A range of length is indicated. The range is determined by the ability of the equipment to recover the signal with an attenuated loss. The first number applies to an attenuated loss of not more than 23 dB at one-half the clock frequency. The second number assumes the signal can be received with an attenuation loss of 30 dB. Where patching, routing, and

Figure 6-22

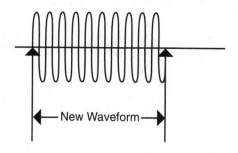

←— New Waveform —→

Figure 6-23

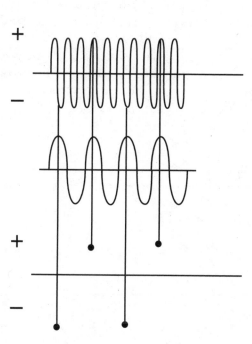

TABLE 6.4

Data Rate and
Cable Distance in
Meters

Part	143 Mbps	177 Mbps	270 Mbps	360 Mbps
8281	325—425	280—385	225—295	190—250
1505A	330—430	290—380	260—340	225—295
1694A	410—540	370—480	310—405	270—350

other equipment is attached, their losses can add to the overall attenuation number, especially at the higher data rates.

The examples in the table should not be considered to show hard-and-fast numbers, but can be used as guidelines to appropriate cable length.

To determine the bandwidth of digital video signals, which are *non-return-to-zero* or *NRZ* systems, a rough conversion of 2 bits = 1 Hz is appropriate, as shown in Table 6.5.

Structural Return Loss

In any coaxial cable, variations in impedance are caused by center conductor migration, variations in the diameter of the dielectric, and braid coverage positioning. There also are variations in velocity of propagation caused by the consistency with which the plastic is foamed. All of these variations in the structure of the cable, and many others, tend to make signals reflect back toward the source rather than to the destination. This effect is known as *standing wave ratio* (SWR), *voltage standing wave*

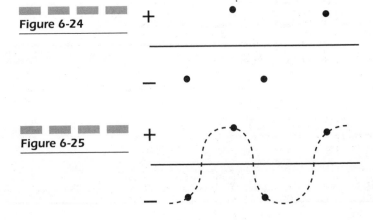

Figure 6-24

Figure 6-25

TABLE 6.5

Data Rate (Mbps)	Equivalent Bandwidth (MHz) and Clock Frequency
143	71.5
177	88.5
270	135
360	180

ratio (VSWR), or *structural return loss* (SRL). At higher frequencies, where slight difference in parameters can have serious results, SRL is a major parameter in determining system performance.

SRL is measured in dB, with the lower number being better than a higher number. Thus a cable that has 30 dB SRL is far superior to a cable with 23 dB SRL. These numbers also apply to connectors, patch panels, splitters, and any other signal-carrying devices. Especially in high-data-rate or high-frequency applications, SRL should be an essential and required number for all parts of a system, not just the cable.

Periodicity

All cables are made by machines. If these machines are not maintained well, it is possible that a minor misadjustment of the machine can cause a major problem in the cable. For instance, if a wheel that feeds the core of a coax is out-of-round, every time it turns it will introduce a minor change in some aspect of the cable (diameter of the dielectric, for instance). While minor by itself, this flaw will be repeated over and over in the cable. The distance between flaws will not be random, but will correspond to some wavelength, and therefore to some frequency and multiples and divisions of that frequency. Because of the aggregate effect of this repeating minor flaw, there can be a substantial spike of SRL at a very narrow band of frequencies. This is called *periodicity*.

The significant SRL spikes shown in Fig. 6-26, if not obviously attached to some major flaw in cable construction, could be attributable to periodicity. Normal attenuation tests will rarely reveal these flaws. Only a full sweep of all frequencies within the desired band will reveal if periodicity exists in any given cable.

Figure 6-26

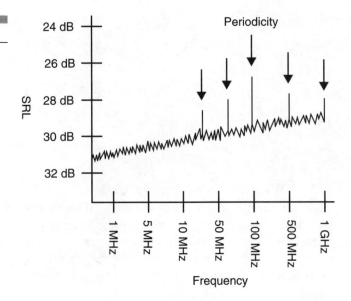

Periodicity is one of the most difficult problems facing manufacturers of wire and cable because an extremely minor flaw, when repeated over and over, can have major consequences. Especially in high-data-rate or high-frequency constructions, only constant maintenance, testing, and measurement of both cable machinery and the cable itself can prevent this problem.

Group Delay

One other factor to be considered for high-data-rate or high-frequency constructions is *group delay*. The effect is that, in a broadband signal, some frequencies will arrive sooner or later than other frequencies. This is caused by changes in the velocity of propagation of the cable compared to frequency. In solid-dielectric cables, the choice of dielectric material is critical because some plastics will vary in their V_p. Polyethylene and Teflon are the most common stable compounds. In foam dielectrics, the consistency of the foaming can have a profound effect on V_p and group delay.

Often, cable manufacturers will specify the V_p tolerance (i.e, a maximum and minimum value). A tight tolerance indicates low group delay. Manufacturers usually can monitor capacitance as the cable is being made. Since capacitance is directly related to dielectric constant, variations can be measured and adjustments made as the cable is being extruded. There is no visual way to tell if any particular piece of cable has a group delay problem. It must be tested with a broadband signal and the results noted.

Embedded Audio

A common practice in digital video is to digitally encode the audio (AES/EBU) and to embed it in the picture. It can be added as lines of data before, or even during, the picture data. This means that the sound will always follow picture. Embedding is an especially effective means for transmitting multiple channels. This is a transmission protocol, however, not an editing protocol; the audio must be decoded to edit, process, or manipulate it.

If digital video with embedded audio is run through video processing, switchers (especially switchers that do picture manipulation), or other active devices, the audio may be interpreted as extraneous noise and the processing may strip it off the picture.

Only by being sure the equipment will not strip off the audio, or by decoding it before picture processing and adding it back afterwards, can you be assured of not losing the audio.

Digital Video Patchbays

If you've gone to the trouble of installing cable for digital video, you might as well do the same for your patchbays. Unlike AES/EBU digital audio, digital video patchbays are readily available.

The two key factors are impedance and bandwidth. The jacks (and patch cords) should be true 75Ω in and out. Many of the older patch bays are 50Ω. If it doesn't say 75Ω in the manufacturer's literature, act with caution. Digital video signals can have extremely broad bandwidth; clock frequencies up to 135 MHz are common. Therefore the bandwidth of a jack or patchcord should be at least 200 MHz, if not more. It's even better if the manufacturer gives you SRL numbers. These indicate the reflections at various frequencies and truly indicate the performance of each jack.

7

Multimedia
High-Data-Rate
Cables

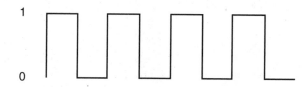

A tremendous amount of work has been done in the last few years on twisted-pair cables in the data world. Most of this has gone unnoticed by the broadcast community. Talk to any computer installer and he'll be talking about Category 3 or 5 twisted pairs. This revolution in wire and cable will affect broadcast very soon. It's the backbone of the multimedia revolution.

What Is Multimedia?

The truthful answer is that nobody really knows what multimedia is, mainly because it is evolving as the technology changes. It is our lack of vision and imagination that prevents us from seeing what is to come. We can, however, make an educated guess as to what and where multimedia will appear.

As the name implies, multimedia seems to be the intersection of a number of separate technologies. The first is publishing, the second is broadcasting, and the third is the telephone. All of these are united by a common technology, the computer.

The first step is to take the information desired and turn it into a digital signal. Once in that form, it can be stored, processed, and transmitted like any other digital signal. The revolution is in getting all these different digital signals into a common frame of reference.

Already there are little fingers of interactivity pointing the way. Publishing and the telephone companies have spawned the fax machine. Get a fax card for a computer to send and receive faxes. Once a fax is received, it be can printed, stored, or retransmitted. There are now sound and video grabber cards that will process audio or video information and allow them to be stored, manipulated, or played by a computer, or even transmitted through a network to other computers.

Within a few years, no doubt, you will buy a box that will do word processing and spreadsheets, show TV programs with unparalleled picture and sound quality, allow you to call up programs (audio, video, or data) at any time and manipulate them any way desired, allow remote shopping, order airline tickets, pay bills, make stock transactions and other consumer and financial services, do research in any library on earth, and allow the presentation of data (including audio and video) for retrieval and consumption by others.

In fact, I will go out on a limb and predict that the major source of distributed public entertainment will be home-generated programming! We will entertain ourselves. The stars of tomorrow are sitting in their living rooms today, and will be famous without leaving home!

Data Delivery

Like the confusion about these convergent technologies, there is confusion about delivery systems and the wire and cable that will deliver it.

For audio and video in its source form, you will use standard analog audio and analog video cable. Once it is digitized (if it is a standalone system such as broadcast-quality digital video or AES/EBU digital audio), it will have its own special family of cables to handle its particular needs.

Once that digital signal has been manipulated to come to the common ground of multimedia, it will also have common cables. These will be data cables because, by that point, all these signals will be just data. There are three types of cables used for data: coaxial, twisted-pair, and fiber optic. Surprisingly, the one that is dying is coax cable. The other two are doing fine!

SCSI

SCSI means *Small Computer System Interface*. It is a system that allows the transfer of data down stranded twisted pairs. The data protocol is for 8-bit words at a maximum of 5 Mbps. It is a multipair format utilizing 25-, 34-, and 50-pair cable with an overall shield. In a balanced line configuration, these are specified as 120Ω impedance. Unbalanced, they are 80Ω. These stranded, low-capacitance designs attempt to overcome the natural limitations of the cable by using multiple pairs (much like parallel digital video). The more pairs, the higher the data rate. Connectors with this many pins are time-consuming to connectorize and bulky when finished.

The main advantage to SCSI is the simplicity of the driving and receiving architecture. However, you can only go 6 m (20 ft) in the unbalanced mode and 25 m (81 ft) in the balanced mode before data is unreadable. This is not appropriate for large installations. And, since this standard was started when 4 Mbps was considered "fast," it is rapidly becoming outdated.

To overcome some of the speed limitations, there is fast SCSI, wide SCSI, and even fast/wide SCSI. "Fast" means a 10 Mbps data rate (doubling the data rate). "Wide" means 16-bit words instead of 8-bit (again, doubling the data rate). The combination of fast/wide allows for data rates up to 40 Mbps. However, this is still a far cry from 143 Mbps or 270 Mbps digital video data rates. Even at 40 Mbps, the attenuation and distance limitations of the cable are largely ignored, and the maximum length of run of cable becomes even more limited.

RISC

RISC means *reduced instruction set computing,* a somewhat newer approach to microprocessor architecture often used in the desktop computers called *workstations.* Some manufacturers of RISC devices, notably IBM in their RS/6000 series, have also mandated peripheral interconnection and communication standards for these machines. IBM's is a protocol-based standard relying on moderate-quality stranded 100 and 120Ω shielded pairs for data transfer.

Unlike SCSI, the RISC standard is based on three twisted pairs and RJ-45 jacks and plugs. These plugs and jacks are slightly larger versions of the same clear plastic plugs used in telephones. Even high-quality, high-data-rate RJ45s are very inexpensive. They are installed by insulation displacement; the wires are stuck into slots in the connectors and a special tool squeezes the wires onto pins, where the insulation is displaced (i.e., cut through) and connection is made. For paired cable, this type of connector is the fastest, simplest, and cheapest.

RISC cable suffers from one inherent flaw, however. The choice of RJ45 as the connector standard means the size of wire that fits into the slots of the connector (and therefore the amount of insulation) is predetermined. Since wire gage and insulation thickness are predetermined, electrical parameters such as capacitance are limited. These limitations dramatically affect data throughput and length of runs.

The only solution to this problem so far is *dual extrusions,* where there are two layers of plastic over the wire. With two layers, the capacitance of a pair is acceptably low (12 to 15 pF/ft), and reasonable run lengths at reasonably high data rates can be achieved. To put on a connector, you strip off only the top insulation layer; the lower layer is the correct dimension to fit into an insulation-displacement RJ-45.

Twinax

Many older data transfer and networking schemes use *twinax*. There is even 124Ω triax, the earliest attempt at balanced-line video. Twinax is still in use for data networking and machine control, especially in IBM mainframe environments. It was the earliest version of impedance-specific twisted pairs. Constructions are available in various impedances from 78 to 200Ω. Impedance tolerance is achieved by "packing" the cable with nonconducting plastic fillers to immobilize the twisted pair and to keep them predictably spaced for low capacitance (7 to 20 pF/ft). For true long-run, high-data-rate performance, however, all these stranded twisted-pair solutions pale compared to coax and solid twisted-pair cable.

Data and Coax

Coax has been the cable of choice since the dawn of the computer. Many of the reasons are obvious. Cable and connectors are very available. It is easy to connectorize. It has very stable impedance measurements, compared to twisted-pair cable, for good high-frequency performance. The most common is Ethernet, which is a 50Ω standard. There are versions of 50Ω cable for Ethernet, specifically a large-size, low-loss, long-run, quad-shield trunk cable called *Thicknet*, a low-loss RG-58 version called *Thinnet*, and a short-run lower-quality version commonly nicknamed *Cheapnet* (which uses generic RG-58 as the cable).

In networks such as Ethernet, the difference in quality between common RG-58 and true Thinnet can be dramatic. It may become apparent, however, only when one more node is added to the network and it doesn't work. Naturally the quality of the generic RG-58 is a major factor, because SRL and other cable factors will determine the quality of transmission.

There are networking schemes for shipping audio and video down Ethernet. These provide only low-data-rate, low-quality images. They are fine for pictures on the desktop, but none of them even approach broadcast quality.

Solid-Wire Pairs

Solid-wire twisted pairs are a known cable construction. They have been around since the dawn of the telephone. Any phone installer can install these cables. In many instances they even use the same color code as standard telephone cables.

Unshielded twisted-pairs (UTP), and the less common shielded twisted-pairs (STP) have many advantages over coax. They are cheap, in some cases much cheaper. They are balanced line and thus have common-mode noise rejection superior to any coaxial cable. They are available in bundles, up to 25 pairs or more, depending on the data speed desired.

Unlike SCSI or parallel digital video, each pair contains the full bandwidth. Interactions between pairs (such as crosstalk and capacitance) increase with multiple pairs, so higher pair counts generally offer lower bandwidth and lower data rates. Each pair is equivalent to a coax, so the space-saving, installation, and handling ease of twisted pairs is vastly better. Installation costs of twisted pairs, even exotic high-data-rate twisted pairs, are orders of magnitude lower than coax cable in comparable installations.

How High Is "High Data Rate"?

As enumerated in Table 7.1, there currently are five categories of data speed.

The TIA and EIA are two organizations that determine the specs for high-data-rate twisted pairs. They are the Transmission Industry Association and the Electronic Industry Association. More TIA/EIA specifications are in the "Notes and Comments" section (Appendix A) of this book. You can obtain a complete copy of the TIA/EIA 568A standards

TABLE 7.1

Category	Medium	Impedance	Bandwidth	Data Speed
1	UTP-POTS	unspecified	unspecified	20 kbps
2	UTP	unspecified	unspecified	4 Mbps
3	UTP/STP	$100\Omega \pm 15\Omega$	16 MHz	10 Mbps
4	UTP/STP	$100\Omega \pm 15\Omega$	20 MHz	16 Mbps
5	UTP/STP	$100\Omega \pm 15\Omega$	100 MHz	100 Mbps

document by calling (800) 854-7179. Also available are the Technical Systems Bulletins TSB-36 (about Category 3-4-5), and TSB-40 (about connectors, patch cables, and recommended wiring practices).

POTS is "plain old telephone service," the standard twisted pairs that come into today's home (see Appendix A, item 18). These cables were never intended for high-speed data service, and TIA/EIA gives them no such specifications. As we will see later, however, many companies are eyeing POTS lines with schemes to use them for high-speed data anyway!

Most current phone wiring is Category 3, a conscious attempt to future-proof installations. The only question is how high should the specs be? How far into the future should you prepare? Many installers are putting in Category 5 now because it is the highest data rate UTP/STP available. Often the installer will put in Category 3 for phone and Category 5 for data. The minor amount of money saved (most of the cost is for labor, not wire) could lead to problems later, however.

Let's say the boss's office was wired up with Category 3 to the phone and Category 5 to the computer. Let's suppose that the boss gets tired of the view out the window. He wants to look out the other window. You can move the furniture around, but not switch the phone and computer. If only everything were wired up Category 5!

For a minor cost increase, the same cable could be used everywhere. In the phone closet, where most likely there is a data patch panel, you can accomplish the move merely by switching the patch cords for the phone and computer.

Still the question remains: How fast a data rate is fast? Broadcasters currently have higher data rates than virtually any other end user. Rates of 270 Mbps (NTSC) or 1.485 Gbps (HDTV) are far faster than even the most cutting-edge network data systems (Appendix A, "Notes and Comments," item 29).

Asynchronous Transfer Mode

At 155 Mbps, *Asynchronous Transfer Mode* (or ATM) currently is the fastest data transfer system. Considerable work is being done on putting video to the desktop down ATM. However, there is a misconception about ATM and video.

Digitized composite NTSC is only 143 Mbps. While this could handle a substantial amount of data, it cannot handle broadcast-quality video. The word size and error correction scheme of ATM and digital NTSC are vastly different. Sony has attempted to use ATM for broadcast video and has abandoned it because of these limitations. They have opted for a proprietary networking scheme for digital video, which also might not be much of an industry-wide solution.

To put 143 Mbps onto ATM at 155 Mbps means, for one thing, that there is only one user on the line. And while there are also higher ATM data rates than this, they are all fiber-based, and there are no plans to abandon coax and go all-fiber in broadcast. Conversion from electronic to optical, at every input and output, is alone a formidable expense.

The video currently seen at the desktop is nowhere near broadcast quality. At best, it is of the 6—10 Mbps quality; often it is much lower (T-1 or less) "near-VHS quality."

At the same time there is the conflict in presentation between progressive scan (which is how computer monitors display their information) and interlaced (the home television receiver method). There are other decisions to be made that are more basic than wire selection or even appropriate data rates (Appendix A, item 29).

Data Rates

Much work is being done on *multilevel coding.* These are digital schemes which, instead of a "zero and one" scheme, have in-between levels that can be encoded and decoded.

This approach has two advantages. First, you can get more data in the same space. Second, multilevel coding reduces the apparent clock frequency, thus reducing the critical specifications on the cable. You can go farther on high-data-rate cable, or use lower-speed cables for standard distances

The multilevel transmit and receive boxes can be complex and expensive, and this protocol is only a transmission medium. No system is set up to process video or crunch numbers in this format. This makes conversion and reconversion an expensive proposition.

Testing

There is continuing controversy in the area of specifications and measurements, especially with high-data-rate cables.

Since 100 Mbps twisted pairs are cutting-edge, much work is being done on testing and verifying system performance. This is very difficult without the right tools. Merely testing continuity with an ohmmeter tells nothing about crosstalk, impedance variations (which lead to increased structural return loss and signal reflection), connector performance, velocity of propagation, or any other performance factor, including resistance variations (resistance unbalance).

Some manufacturers take a four-pair cable, test all four pairs, and use the best one to determine specifications. This can lead to performance claims that cannot be substantiated in the field. Some just quote the TIA/EIA specs, so there is no way to tell how their cables actually perform. Still others test their cables in nonstandard ways to arrive at better test numbers than "typical" values. The only true way to know is to actually test the cable in the field.

Many Category 5 testers have a chip with the manufacturer's data burned in. If that data is faulty, so is every reading. How the data for the specification was arrived at also is controversial. Some manufacturers measure at "eleven critical frequencies" (which go unspecified), while others measure many more points, sometimes hundreds of frequencies.

If the latter manufacturers use their worst-case pair and measure hundreds of frequencies, this data will look substantially worse than the competition using best-case pair and eleven frequencies—even though in actual performance the reverse might be true.

In theory, of course, you could phone each manufacturer before buying the cable and find out how the specifications in the data sheet were arrived at, assuming the manufacturer would share this information. There is only one way to tell the performance limitations of any cable, however, and that is take accurate measurements of the cable yourself. To do this, you must manually calibrate the test instrument for each type of cable.

A problem exists with handheld testers, some of which lack the precision necessary for true Category 5 testing. Some are as much as 30 percent off actual test values. In manual testing, a known sample of a known exact length is entered as a baseline set of parameters in a tester. The installed cables are then compared to it. This gives truly accurate performance specifications.

There is a serious drive within the data community to do *link testing.* This is a philosophical shift in testing, which recognizes that it's not just the cable, or the outlet/connector, or the patch panel or the hub, or any other single part, that has specifications critical to overall performance. It is the system as a whole that should be tested. All the pieces, when linked together, will show up aberrations of dueling algorithms, error sensing/correction, and other parameters that would have been missed completely had each piece been tested separately.

Concatenation, also called *dueling algorithms*, occur when signals are successively digitized and then converted back to analog, only to be digitized again and converted back again. The subtle changes that build up in the resulting analog version can be passed on in ways that are impossible to predict when digitized a second or third time. It is safe to say, however, that very little or none of this is a cable problem. It is almost certainly a function of the conversion process and the error sensing and correction at each stage of that process.

Where signals are purely data, and never converted to or from analog, link testing attempts to show the subtle interaction between various pieces in the data chain: hubs, routers, concentrators, error sensing/correction and, of course, cable.

Variations in impedance, with SRL and attenuation losses (especially in long runs), can affect error sensing and correction when the signal is too low to be interpreted accurately.

TIA/EIA 568A has an informative supplement or "Annex" with the specifications for UTP link testing. As these systems become more and more embedded in the digital audio and video worlds, there will be link specifications for these systems too—although there are currently no systems or specifications available.

Category Cables and AES/EBU

How category testing or the published cable specifications apply to digital audio is sometimes difficult to determine, since data specialists are concerned with 100 Mbps data rates, and audio engineers are only interested in a 3 Mbps data rate.

As a rule of thumb, stick with a manufacturer who speaks fluent AES/EBU and TIA/EIA. They can help sort out the mess instead of complicating things. They should provide attenuation and crosstalk numbers at some bandwith close to AES/EBU (such as 4 MHz) This, together with the capacitance and impedance of the cable, should give a clear picture of the appropriate use for the cable in question.

UTP Installation Practices

Performance of any data system is affected as much by installation practices as by selection of wire and cable. It is very easy for an installer to take 100 MHz Category 5 cable, or any other grade of cable, and turn it into cable suitable only for telephone service. Here are some of the guidelines to help maintain performance.

- **Use proper connecting hardware** If installing Category 5, all the hardware, jacks, plugs, cross-connects, patch panels, and patch cords also should be Category 5 rated. Make sure the rating is written on the item, in a catalog, or in a data sheet. Performance cannot be determined with the naked eye!

- **Test the equipment and analyzers** If installing Category 5, measure Category 5. Make sure the test equipment is set up for the appropriate bandwidth/data rate. If there is a preprogrammed section containing ostensible values for different wires made by different manufacturers, be especially cautious. Remember that the best way to test a network is to take a known-good sample of specific length, test it, and then run it against the rest of the installation. Nothing is as good as direct comparison.

- **Eliminate tension stress** Be gentle. The higher the data rate, the more the impedance and SRL are affected by excessive pulling. Strong pulling also forces the pairs to shift and "nest" with each other, increasing crosstalk. Pulling ultimately will elongate the conductors, causing resistance unbalance and almost certain failure of that cable at high data rates.

- **Avoid tightly cinched bundles** To change the impedance of a twisted-pair cable, just squeeze it together. If cable ties are necessary, leave them as open loops. Do not cinch them tightly (as might happen with a tie gun) to harness them down.

- **Keep kinks out** In a twisted-pair cable with a sharp kink, the impedance and SRL at that point can be severely degraded, especially in high-data-rate cable. However, bonded pairs do unusually well in such a situation.

- **Strip only as much as necessary** Jacket stripping is determined by the requirements of connectors and hardware. The rule of thumb is no more than an inch. Six inches will almost guarantee a failure of some sort.

- **Reduce untwisting of pairs** Since it's the twisting that helped determine impedance in the first place, untwist the pairs as little as possible, no greater than one-half inch for Category 5.

- **Make no 180° bends** Again, be gentle. Bend radii ideally should be less than 4 times the cable diameter for horizontal runs, and never more than 10 times the diameter.

- **Stagger the cable ties** If a number of cable ties are evenly spaced down a cable, an SRL spike at some frequency will probably occur. That frequency will be determined by the wavelength corresponding to the spacing of the ties. Try to randomize their spacing.

Observing these cautions, and using a nominal amount of plain common sense, should ensure that any problem that arises is not in your cabling.

Fiber Optic Cable

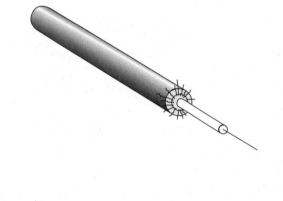

When and When Not to Use Fiber

Fiber optics and data have been around for over 30 years, but fiber optics and broadcast are new partners. Therefore, there is a lot of "fear of fiber" among audio and video people. After all, it cannot be soldered or crimped and it can't be tested with an ohmmeter. Fiber is not wire.

So when would you use fiber optic cable? There are two answers: when it is necessary and when it makes sense. It is necessary:

■ If you have a box that has fiber connections to it.

■ If the phone company installs fiber to a location and an extension to that fiber cable is desired, while still staying in the fiber mode.

■ When a customer demands fiber as part of an installation.

Fiber makes sense:

■ For very long runs—anything from hundreds or thousands of feet, to miles or even hundreds of miles (depending on the kind of fiber).

■ In very noisy environments with electromagnetic or radio frequency noise. Fiber is immune to every kind of electromagnetic noise, even lightning.

■ In exposed environments. Fiber is immune to oxidization and corrosion, which are potential problems with copper cables.

■ When the size or weight of copper cables are major considerations. Fiber can be extremely small and lightweight, depending upon how it's bundled.

■ For very high data rates or very high-bandwidth analog signals. "High" means essentially unlimited! Certain types of fiber have been tested in the laboratory up to 40 GHz of bandwidth, with no limit in sight.

Many installers are putting in fiber and leaving it "dark" (i.e., unconnected). If there is no doubt that an installation will go to fiber within the next 10 years, then it may make sense to install fiber now. The fiber itself is fairly inexpensive and the labor to put it in place will be about the same as installing copper. Some manufacturers even make copper/fiber hybrid cables for exactly this purpose. Category 5 pairs and 62.5 μm (micron) fibers are a common combination.

There are definitely times when fiber is not appropriate. To use it because it's "in," or to make an installation look "cutting-edge," is an expensive proposition. While the fiber itself is not expensive, each piece of equipment that does not have optical inputs or outputs will have to be converted. The boxes that convert from electrical to optical are not cheap (though they're getting cheaper), and you will end up paying much more than you would for a copper installation of similar quality.

Fiber Basics

There are three kinds of fiber, one made of plastic and two made of glass. Plastic fiber is of large diameter (by fiber standards), usually 900 µm (microns). It is generally made to conduct visible light. Wavelengths of visible light are short compared to the infrared waves used in other fibers. That fact, coupled with the "huge" diameter of the fiber, means the light is bouncing all around as it travels down the bore. Much of the light is reflected back or dispersed; only a small percentage is left after relatively short distances. This is why plastic fiber usually can't conduct light more than 20—30 ft without excessive losses. The advantages of plastic fiber are that it's cheap, and that connectors are easy to put on because of the large diameter.

Glass fiber is very different from plastic. Figure 8-1 shows what a single glass fiber looks like.

You might see catalogs specifying glass fiber as "62.5/125/900" or similar nomenclature. These numbers refer to the diameter in microns of the core (62.5 µm), with the cladding added (125 µm), and with the coating added (900 µm). The core is the actual fiber. The cladding is made of the same material as the core and is designed to lower the refractive index, that is, help keep the light in the fiber and not let it out to bounce around. The coating is multiple layers of plastic, which helps protect the fiber.

Multimode Fiber

Glass fiber comes in two versions: multimode and single-mode. *Multimode* refers to the fact that the light has multiple pathways or *modes* to travel down the single glass fiber. As in plastic fiber, this limits the

Figure 8-1

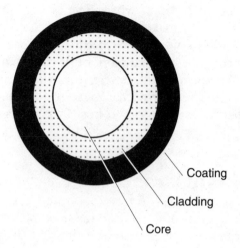

Coating

Cladding

Core

distance the signal can go before reflections drop the light level below recoverability. Glass multimode fiber can go thousands of feet before reflection use up the signal, however, much farther than plastic fiber. Because glass fiber is so much smaller than plastic, its aperture (light-gathering size) is equally small.

Multimode fiber is most commonly found in "short-haul" applications of 1000 ft or less, in installations such as building or office cabling. It is available in many different sizes of fiber: 50, 62.5, 85, 100, and 200 μm, plus some oddball sizes. The most common sizes are 50, 62.5, and 100 μm, with 62.5 μm the most common. Note that 62.5 is over 14 times smaller than plastic fiber and, like almost all glass fiber, uses ultraviolet light. As a result, it can handle much more data and the signal can go much farther.

Multimode connectors are cheap and easily available. They can be easily installed, even in the field. Of course, the installation process is very different from copper cable!

Glass fiber works at two light wavelength openings or "windows." The wavelengths are measured in *nanometers* (billionths of a meter, abbreviated *nm*). These two windows can carry signals simultaneously, so the total bandwidth of a fiber is the sum of the bandwidth for both windows. For multimode fiber, the windows are 850 nm and 1300 nm. Because the wavelengths of 850 and 1300 nm behave differently in different fiber sizes, these windows vary significantly in the amount of bandwidth they can handle. In 50 μm fiber, for instance, the two windows are the same; at 100 μm, the two windows are 50 percent different from one other.

62.5 μm fiber has a combined bandwidth of 660 MHz/km. There are still some, however, who believe that super-wide bandwidth, single-mode fiber must be used for high-data-rate digital video. This is not true. Multimode fiber covers every current digital broadcast protocol, including uncompressed 1.5 Gbps HDTV! The drivers and receivers that will work at these data rates are expensive, but not the fiber.

Step Index and Graded Index

In multimode fiber, where the transmission characteristics between core and cladding differ dramatically, this construction is said to be a *step index* design (Fig. 8-2). The propagation of the light is at much sharper angles because the cladding is, by design, a much poorer light conductor.

Much more common is *graded index,* where there is a less abrupt change in transmission characteristics between the core and cladding (Fig. 8-3). The propagation of light is more wavelike as the cladding is a poorer light conductor, not an abrupt change.

What Size to Use?

The 62.5 μm size has won out over the other sizes because it comes up with the best overall specifications. Table 8.1 shows the three most common sizes of multimode, with specifications and a grade (A, B, C) for that specification.

When considering buying equipment that uses fiber, be aware that 50 and 100 μm, while still common, are getting less and less so; 62.5 μm has all but won the "battle of the microns" when it comes to multimode

Figure 8-2

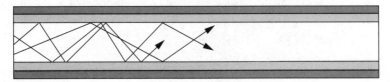

Step index

Figure 8-3

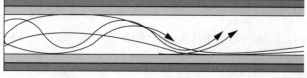

Graded index

TABLE 8.1

Specification	50 μm	62.5 μm	100 μm
Numerical Aperture Light gathering	0.2 (C)	0.275 (B)	0.29 (A)
Attenuation dB/km	3.0 (A)	3.5 (B+)	5 (C)
Bandwidth MHz at 850/1300 nm/km	500/500 (A)	160/500 (B)	100/200 (D)
Handling Ease of manufacturing	(C)	(A)	(C)
Availability Most commonly stocked	(C)	(A)	(C)
Connectorizing Ease of attaching connector	(B)	(B+)	(A)
Overall Grade	(C+)	(A—)	(B—)

fiber. So if someone offers a deal on some other size than 62.5 μm, ask tough questions. Unless this is a one-off, standalone installation that will never be modified or expanded, in a few years it might be hard to get fiber and connectors for that other size.

Single-Mode Fiber

Single-mode is very different from multimode (Fig. 8-4). This is the fiber used by the Baby Bells, AT&T, Sprint, MCI, etc. This is the super-wide-bandwidth, long-haul cable (see Appendix A, "Notes and Comments," item 24). To send dozens of gigabits hundreds (even thousands) of miles, single-mode fiber is the best choice.

Single-mode is very, very small, usually from 5 to 10 μm. This small size means that the light waves have very little room to bounce around; there is very little reflection and cancellation. It also means that its aperture is very small and it requires a very powerful source (such as a laser) to go long distances. The very small size also makes it difficult to connectorize. In fact, a microscope is required just to see the fiber. Because of this, the connectors can be very expensive and very difficult to attach, especially in the field. It is not unusual for an installer to do the same single-mode connector a half-dozen times before it's right.

Like multimode, single-mode has dual windows, but at 1300 and 1550 nm. There is also *dispersion-shifted* fiber, where the light is concentrated in the 1550 nm window, but this is unusual for normal commercial installations.

Loose and Tight Tubes

Glass fiber comes in two types of package, *loose-tube* and *tight-tube* (Fig. 8-5).

People are always surprised to see how flexible yet rugged fiber optic cable is. And yet it is glass! What ruins fiber are usually temperature fluctuations and water. Severe temperature changes can result in the formation of microcracks, which increase the attenuation of the desired signal until it is unrecoverable. If water seeps into a fiber optic cable and then freezes, it can break the fiber. One solution is to use a *loose tube* (also called a *loose buffer*) around the fiber. Since any jacket material will have a different *coefficient of expansion* (it will expand or contract at a different rate than the glass), the loose tube allows expansion or contraction without affecting cable performance. Any cable intended for outdoors should be loose tube.

A *tight tube* (also called a *tight buffer*) would break the fiber if it were subjected to temperature extremes. That is why "outdoor" tight-tube constructions need to be protected from the elements, such as by conduit. Most tight-tube use is in controlled environments such as offices, buildings, or factories. There the tight tube's advantages of ruggedness and low cost can be more important. Most loose-tube is used in *interbuilding trunk cable*, that is, multifiber, long-run constructions between buildings.

In the office or studio environment, you rarely need more than a few fibers going to any particular box. With interbuilding trunk cable, how-

Figure 8-4

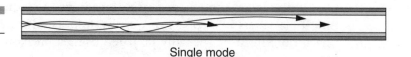

Single mode

Figure 8-5

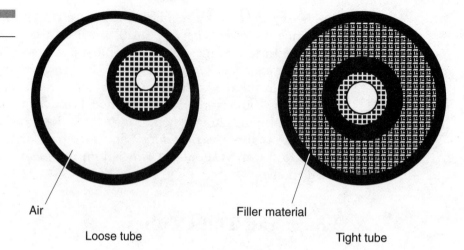

Air

Loose tube

Filler material

Tight tube

ever, dozens or even hundreds of fibers might be required. Because of these two contradictory styles, fiber optic cable is available in two styles: *single fiber per tube* and *multiple fibers per tube.*

Fibers per Tube

One tube for each fiber is more expensive, but much easier to use. Individual tubes allow identification of each fiber either by tube color, numbering, or some other means. When connectors are put on the fiber, they attach to the tube (for strength) as well as to the fiber.

Multiple-fiber tubes can have two to six fibers per tube, or occasionally more. These can be bundled in groups up to 240 fibers total per cable, or even more in specialized constructions. Usually loose-tube constructions of up to six fibers total have a single fiber per tube; after that, there are multiple fibers per tube.

Multiple-fiber constructions are much less expensive, easier to manufacture, and have smaller overall diameters. For very long runs (such as interbuilding or intercity runs), these savings can be a major factor. In those cases, added expense of preparation for connectorization at each end is minor.

On the other hand, if dozens or hundreds of connectors are required, multiple fibers per tube can quickly lose its advantages to labor cost. Each fiber must be specially prepared using a breakout kit, a very laborious project.

Multiple fibers per tube therefore are most often seen (and most cost-effective) in long runs, while networks within buildings most often use single fiber per tube.

Breakout Kits

More than one fiber per tube requires that a *breakout kit* be bought with the fiber. This kit contains heat-shrink tubing, PVC tubing, and possibly some strength members, to allow the preparation of each fiber for connectorization.

Many thin, tight-tube, office-style fiber cables also require a breakout kit because the existing jackets and fiber are not strong enough to allow a connector to be attached.

Strength Members

While fiber can take a considerable amount of bending, its one failing is tension. The glass cannot elongate (lengthen) like copper before it breaks. To help the fiber survive the rigors of pulling and other elongation handling, strength members are put inside the tube with the fiber. The most common of these are aramid fibers.

You might also see steel or fiberglass epoxy rod used as a strength member. While some of these materials might have minor advantages over aramid, they also have many drawbacks. Aramid's breaking strength is almost twice that of steel or fiberglass epoxy rod. Steel weighs almost six times as much, but it has one-third the elongation. Steel is not very flexible and, if you are using fiber to get away from metal (i.e., lightning protection or RFI/EMI conduction), steel is not a good choice. The best compromise is aramid fibers.

Table 8.2 summarizes the features of several types of strength members.

Preparation and Connectorization

Most fiber is prepared in generally the same way. First the cable is cut to the appropriate length, then it is stripped to reveal the coated-and-clad fiber. The fiber is *cleaved* with a special cutter that gives a flat face. Then

TABLE 8.2

	Diameter (in)	Weight (lb/1000 ft)	Elongation before breaking	Tensile strength	Conductive	Flexibility
Aramid fiber	0.093	1.8	2.4%	944 lb	No	Excellent
Fiberglass Epoxy Rod	0.045	1.4	3.5%	480 lb	No	Good
Steel	0.062	7.5	0.7%	480 lb	Yes	Poor

the connector is glued on. The connector is placed in a special holder that holds it vertically. It is then ground by hand on a special grinding cloth to get a perfect face.

All fiber assemblies should be tested after they are made, especially single-mode assemblies. There is no visual way to tell whether the fiber, the connector, or the assembly is correct without testing. Fortunately, testers are common and available.

Doing It Yourself

If you are planning to do your own fiber connectors, you will need to take a class in connectorizing fiber; buy the appropriate stripping, cleaving, and grinding equipment; and buy a tester that will test the assemblies you have made. And if you're set on doing single-mode, you also will need a special microscope and tools to do that assembly. Then, of course, you will need a class or series of classes to learn how to use all this equipment. Unless you are planning to make a career in fiber (or at least use this knowledge on a consistent basis), this may not be a good use of your time and money.

Using a Fiber Assembly House

There are specialized assembly houses or value-added distributors who deal only in fiber assemblies. Some even teach classes in fiber preparation and assembly (Appendix A, item 26). Because this is all they do, they can often make them faster and cheaper than if you did it yourself. Any fiber manufacturer can provide some names and phone numbers or check the Yellow Pages.

Fiber Fire Ratings

For fiber optic cable run inside a building, the cable often must be fire rated. What rating it should have, if any, is the decision of the local building inspector, fire marshal, or other permit official. Most often they follow the National Electrical Code (Appendix A, "Notes and Comments," item 3). The NEC is a voluntary code. Some inspectors go 'way beyond the NEC with their requirements, while others don't require anything at all. The key thing to do is to ask about fire ratings before buying or installing materials.

For fiber there are two basic fire ratings, *riser* and *plenum*. In other types of wire and cable, riser is a rare rating. Not so in fiber. In fact, riser-rated fiber is the "standard" fiber. With riser-rated fiber you can go anywhere except a plenum area. Plenum-rated fiber is more expensive than riser-rated fiber.

When installed in a harsh environment such as direct burial, fiber (or any other cable, for that matter) can be jacketed for heavy duty. There are three basic types: black high-density polyethylene, aluminum armor, and steel armor.

In outdoor installations, such as runs between buildings, the inspectors no longer care what the fire rating is, so most interbuilding trunk cable is not rated for flammability. In those cases ruggedness, water-blocking, and other factors may be much more important.

Ruggedizing can take two basic forms: heavy jackets of strong plastic, usually black high-density polyethylene, or armor. Armor can be aluminum or steel and usually is of interlocking construction, like the BX cable used in electrical wiring. Interlocking armor gives the finished assembly some flexibility, so it can be wound on and off a roll. A heavy-duty jacket usually covers the armor. Extruded armor is a solid corrugated piece of metal, stronger but less flexible than interlocking armor.

Waterblocking is a layer of gel between the armor and the fiber. An alternate construction is a gel-filled tube that contains the fibers. Water is the deadly enemy of fiber. If water gets into or next to fiber, its coefficient of expansion is very different from the fiber. It can put microcracks in the fiber. These are often detected only as increased signal attenuation, which gets worse and worse over time until the fiber is useless. If the water freezes, it can break the fiber completely.

For this reason, interbuilding trunk cable will usually have its fiber in loose tubes (loose buffer). Then the expansion and contraction of the tube due to temperature will not put microcracks in the fiber. If tight-buffer is used outdoors, it must be protected in some way, such as in a tray or conduit. In areas with wide temperature variations, tight-tube should not be used for very long runs.

Connectors

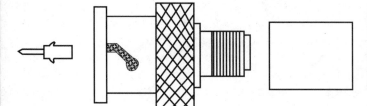

Audio Connectors

Audio connectors come in a dizzying variety of types. Many connection systems don't even require a connector per se, such as crunch blocks and insulation-displacement systems. Adding to the confusion is the difference between balanced and unbalanced connector schemes. With "semiprofessional" audio equipment, more and more unbalanced devices have appeared on the market. Figure 9-1 shows the two most common unbalanced audio connectors.

It is too bad that the phone plug and phono plug have names that are so similar. Be sure you order the right one.

The quality of these connectors can range all over the board, from cheap offshore knockoffs to serious, high-quality connectors. Despite the possibility of high-quality construction, however, these really cannot be considered professional connectors because they are unbalanced. Still, it is amazing to see how often they show up, especially the ubiquitous RCA (or *phono*) plug. Figure 9-2 shows what these connectors look like when taken apart.

The first indication of quality is the materials used. Plastic shells are easily broken and offer no shielding. They are best avoided. There should be a tube of cardboard or plastic inserted in the metal shell. Strangely, the cardboard usually indicates a higher-quality connector: it cannot melt when in contact with hot solder.

There are many connectors, especially the RCA, that come gold-plated. This is no indication of quality. The gold can be so thin on the connectors that one or two insertions will wipe it away from a critical area. If you will be using these connectors in the field, or where corrosion and

Figure 9-1

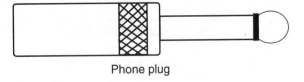

Phone plug

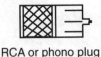

RCA or phono plug

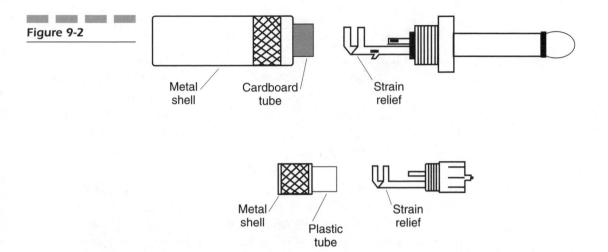

Figure 9-2

Metal shell   Cardboard tube   Strain relief

Metal shell   Plastic tube   Strain relief

oxidation might be a problem, gold-plated connectors (or at least gold-plated center pins) can help lengthen their useful life—but they must have at least 2 microinches of gold (or more) to have any useful aspect.

This rule applies to all connectors: a minimum of 2 microinches of gold on gold-plated surfaces. In fact, if the pins of a connector are plated but not the shell, it is a good indication that this is a professional connector. They're not trying to be flashy and hype you with gold on the outside of the shell, which does nothing to improve performance. Gold pins indicate that the manufacturer is trying to give you better performance. If a manufacturer cannot (or will not) tell you how many microinches of gold is deposited, this is a danger sign as well.

How to Solder

For audio and many control applications, soldering is the most reliable way to make a permanent connection. To solder a connector requires a soldering iron (Fig. 9-3). Do not use a soldering gun or torch. They will melt the parts of the connector as well as the solder. Here are some basic rules for choosing a soldering iron.

1. If you get a simple, plug-in soldering iron, get one no less than 60 watts nor more than 140 watts. (A 100-watt iron is a good compromise).

2. Get one that has removable tips. That way you can change from broad tips for large surfaces, to needle-nose tips for fine work.

3. If you have a choice, get one with a three-pin ac line plug. This will help protect you if you accidentally touch something live, like an exposed ac line.

4. The best irons have adjustable-heat power supplies. These are often called *soldering stations* (Fig. 9-4). These include a sponge in a little tray. Be sure the sponge is moist (not wet) when using it. You can use the sponge to remove excess solder from the tip of the iron.

5. Solder is expensive, so be sure you buy what you need. All solder for electronic use, called *rosin core*, comes with flux (or rosin) mixed in. Solder comes in various mixtures of tin and lead. (There's also very expensive solder containing a percentage of silver.) Standard solder is 60/40, that is 60 percent tin and 40 percent lead. You can get increased tin content, but the solder is harder to use. There are also various diameters. Choose one that will allow you to do small parts in connectors, etc., but not so small that you're paying extra for nothing.

6. Normal soldering procedure is to heat up the physical joint, then feed the solder between the tip of the soldering iron and the surface. If you have insufficient heat, you might form what is called a *cold solder joint*, where the solder has melted but the surface hasn't heated up enough to make it bond.

7. You can often *tin* surfaces or wire, that is, put a little solder on the surface (or the wire) and make sure it melts and adheres. Then, when you go to join the pieces (such as wire to a connector), it will be very easy to melt the solder you have already tinned. Tinning makes soldering go quickly. It is especially recommended when there are plastic parts that might melt, or where you have many wires to connect in a small space.

Making Up Unbalanced Connectors

The simplest connectors are the monaural phone plug and RCA plug. While simplest in construction, they are among the most difficult to solder. These connectors use unbalanced cable, that is, a single conductor with a foil, braid, or serve (spiral) shield. Foil-shielded cable will give you

Figure 9-3

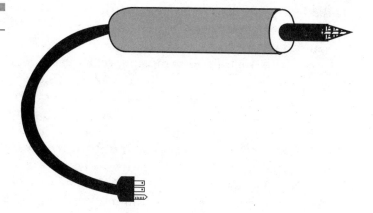

Figure 9-4

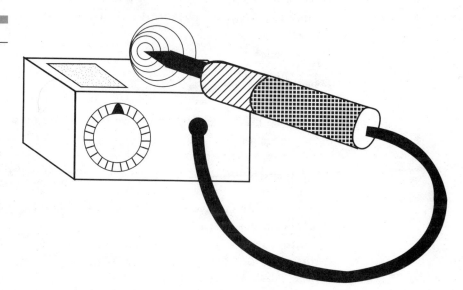

Figure 9-5

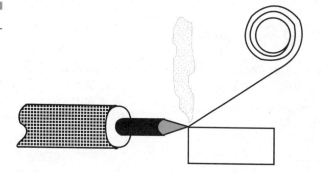

the best performance for the lowest price, but it is hard to find. Braid shielding is superior at analog audio frequencies but is expensive. The vast majority of hi-fi cables have serve shields.

For truly high performance, use coax cables. These are almost always braid-shielded, sometimes foil-and-braid. Their dielectrics give you very high performance, though to some extent they are overkill in analog audio (especially if foamed). Coax is so common that the price, especially in bulk, can be lower than many other single-conductor cables.

Caution! If you use coax for your unbalanced wiring, be sure it is standard or precision video coax, with an all-copper center conductor and a copper braid. *Do not* use CATV "RG-59" cable, which has a copper-clad steel center conductor and aluminum braid. This is a very poor choice for audio performance because it is designed for use starting at 50 MHz!

Here's how to prepare and connect each kind of cable to each connector.

Prepare the Cable

Figure 9-6 shows the first three steps in preparing an unbalancd cable for a connector. First, cut the cable to the correct length. If this cable is installed in a rack or other enclosure, be sure to leave enough for a service loop, if required. Second, slip the metal shell over the cable, making sure it is turned the right way around. Also be sure that the plastic or cardboard tube is included. (Forgetting this step has been the source of much profanity over the years!)

Third, measure dimension X in the connector (from the center pin connection to just in front of the strain relief) and mark that distance on the cable. Alternatively, you can hold the cable against the body of the connector and make a mark on the jacket to indicate the length.

Strip the Jacket

Next, use a cable stripper or single-edge razor blade to cut the jacket to the dimension marked in the previous step (Fig. 9-7). Remove the jacket.

Do not strip so far that the jacket will not connect with the strain relief. Many people assume that the strain relief is the shield connector. This is not so. The purpose of the strain relief is to fold on the jacket and transfer the tension of plugging and unplugging to the jacket, not

Figure 9-6

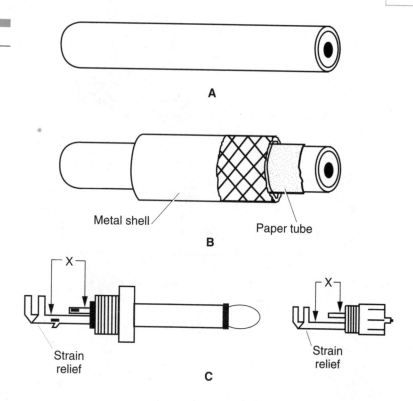

the shield. If you fold on the shield, often the jacket will shrink back, exposing the shield outside the connector. The point of the strain relief is the number-one place where failure occurs inside a connector. The shield alone was never intended to survive being crushed and then flexed; such a connection will have a very short life.

Remove the Shield

Once the jacket is removed, you will come to one of three shields (Fig. 9-8). If it is foil, as in part A of the illustration, trim off the foil. Once you have put a nick in it, it will tear easily. Remove the foil as far back as you can, leaving only the drain wire.

If it is serve (i.e., spiral-wound) as in part B, unspiral the shield and twist the loose wire together. If it is braid (part C), use a pick, a small nail, or similar tool to unlock the braid. This is time-consuming, so don't hurry. It is the price you pay for selecting the best kind of cable.

Figure 9-7

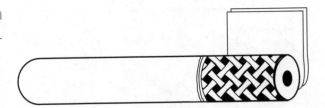

Figure 9-8

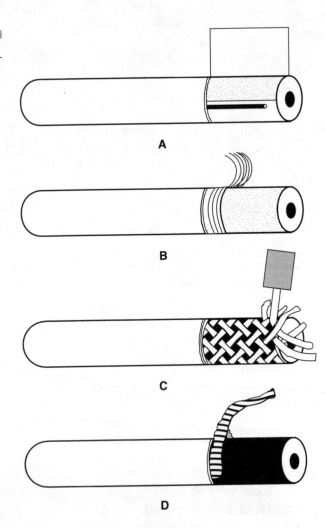

A

B

C

D

Finally, once the braid has been picked apart, twist it together like the serve (part D).

If there is a foil under the braid, remove the foil. If the foil is not bonded, just a nick on the foil will allow you to tear it all the way around the cable. If the foil is bonded (glued) to the cable, it may be impossible to remove. If you're careful when cutting the center conductor, you should be able to leave the bonded foil alone and not have it short out the connector.

Expose the Center Conductor

To reveal the center conductor, determine dimension Y in the connector you are using, as shown in part A of Fig. 9-9. Using your wire stripper or single-edge razor blade, cut that dimension into the center to reveal just that much center conductor as in the lower part of the figure. If it is stranded, twist it tightly together and tin. A tinned stranded center conductor will act pretty much like a solid center conductor.

Up to this point, phone plug and RCA plug wiring are identical.

Figure 9-9

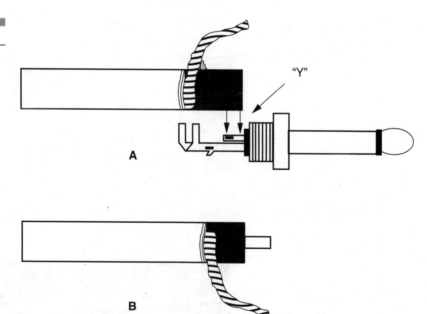

Phone Plug Wiring

If you haven't already, put the metal shell and the plastic or paper tube on the cable. Then turn the cable so the shield or drain wire is facing down. Using pliers, bend the center conductor 90° down (Fig. 9-10).

Make sure that the center conductor fits into the small hole coming from the center of the phone plug, called the *hot pin,* shown in Fig. 9-11. (You'll solder the hot pin last.) Fit the wound braid, wound serve, or drain wire through the ground pin hole and solder it.

It helps to tin the parts of the connector where you will be attaching wire or braid. Unfortunately, it is difficult to tin the connector without filling in the holes you will be using. One easy way is to tin the entire surface you will be using and then suck out the holes with a *solder sucker* (Fig. 9-12). This is a great tool and is highly recommended, especially when you are reusing old connectors.

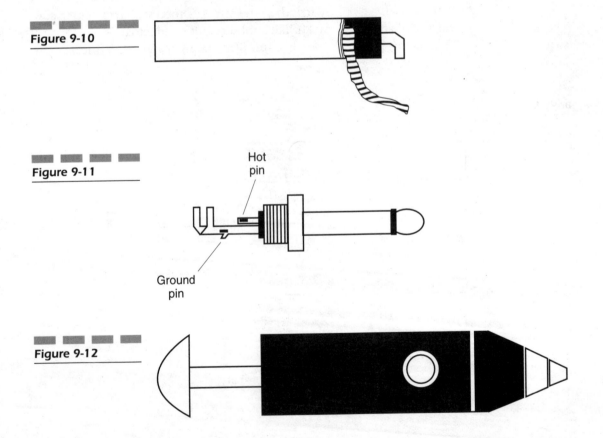

Figure 9-10

Figure 9-11

Hot
pin

Ground
pin

Figure 9-12

Figure 9-13

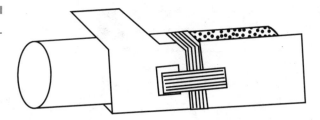

Before you tin the wire, you may notice that you have a problem: You have too much shield to solder. There are two ways to deal with this. The first way is to cut off strands of the shield, which you have combined, until it fits through the hole, and then solder. The problem with this is reduced strength and reduced reliability.

The second way is to divide the strands into three groups. Make the middle group a small enough quantity to feed through the hole. Feed it through the hole. Lay the other pieces of braid across the underside of the connector and solder (Fig. 9-13). If these pieces have all been tinned (including the connector), you can combine all these strands and connect them securely to the ground pin of the connector.

Finally, if it is a stranded center, twist the wires together, tin them and solder the hot pin in place. Soldering it first would have made all the gyrations with the braid shield impossible. Since the hot pin and the center conductor are tinned, it should only take a second to melt the two together.

RCA Plug Wiring

Start with the wire prepared as shown in Fig. 9-14. If the center conductor is stranded, twist it together. Whether it's solid or stranded, tin the center conductor. (You may wish to wait before tinning the shield.) Slide the metal sleeve and plastic or paper tube over the cable.

Position the connector so the center conductor goes into the pin; the strain relief touches the jacket of the cable; and the braid, serve, or drain wire can fold in back of the strain relief.

Insert the center conductor in the connector and position it so that the shield wire is next to the strain relief, but the end of the strain relief is next to the jacket (Fig. 9-15). Using crimpers or pliers, crimp the strain relief to the jacket of the cable.

Fold the shield or ground wire around the strain relief. There is usually a lot more than you need. Cut off the excess. Solder the shield or

Figure 9-14

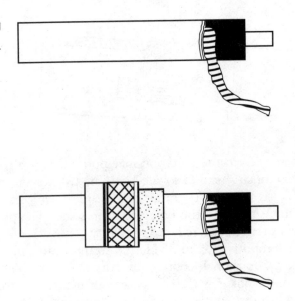

drain wire to the strain relief. Make sure it is flat enough that the tube and shell can slide over the soldered area.

The last step is to solder the center conductor inside the hot pin. It is this step that makes RCA connectors one of the hardest to solder. If you have followed the steps up to now, it should be easy. The trick is what is called the *chimney effect*. That is the same effect that makes smoke rise up a chimney. Heating the center pin will make solder travel up the pin and melt onto the conductor inside (Fig. 9-16).

Unbalanced Wiring with Balanced Wire

If you are doing balanced-line wiring with foil-shielded twisted pairs and you want to wire your unbalanced equipment with the same wire, there are three ways.

- **Cut** Simply cut out one wire of the twisted pair and use the other. Most people cut the black wire and leave the red, just because red is more noticeable, and red indicates the "hot" conductor.
- **Ground** Ground the unused wire at both ends. Again it makes more sense to leave red as the hot and ground black. This has the added advantage of giving a lower-resistance path for

Figure 9-15

Wind shield
Squeeze strain
Relief

Fold against
Strain relief

Solder

Figure 9-16

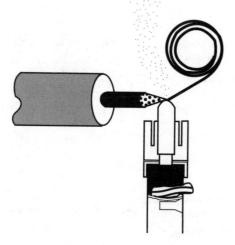

ground—although with high-impedance unbalanced circuits, it probably doesn't make a measurable difference.

■ **Combine** If you strip both the red and black and twist them together, you can put them both in the center pin. This makes the assumption that your pin or solder tab has enough room for both wires. The advantage to this method is that, by terminating a twisted pair onto a single point, you will actually get common-mode noise rejection at that pin. Since most unbalanced hi-fi devices already operate at low levels (i.e., −10 dB), noise reduction is always a good idea.

Connectors for Balanced Line

Audio connectors for balanced line all have (at least) three pins, or multiples that can be divided into groups of three. The most common is the XLR. While XLR originally meant just the version made by Cannon (now ITT Cannon), it is now the generic designation for a family of balanced-line connectors

In all balanced-line connectors (including XLR, TRS, Bantam, DIN, and Tuchel, among others), the three pins connect the two wires of the balanced line and use the third pin for ground. The TRS and Bantam plugs have already been covered in the patch panel sections of Chap. 5 and Chap. 6. There are also two European balanced connectors, Tuchel and DIN. Like the XLR, they are available in various pin combinations, of which three is the most popular for balanced-line applications.

Because of the universal nature of the XLR, and the rarity of other connectors in the USA, we will concentrate on the XLR. Despite the sophistication and relatively high price of an XLR, they are among the easiest connectors to solder and assemble (Fig. 9-17).

When soldering an XLR, have a small cup or container for the parts. When you disassemble them, the parts are easy to lose and difficult to replace.

In a balanced line, if a ground wire accidentally touches one of the balanced-line wires, this would unbalance the line and greatly reduce the noise rejection. It could also allow ground loops (see Chap. 10, "Grounding"), which could create hum and noise. Therefore it is recommended that all ground connections longer than an inch have a piece of plastic sleeving put on them before the wire is soldered into place (Fig. 9-18).

This sleeving is available in a number of diameters, so you may want to experiment with the cable you are using to determine the correct size. If you will be doing very close soldering, you might want to consider Teflon sleeving. It is much more expensive but has a much higher melting point (200°C) than PVC or PE tubing.

XLR-Style Connectors

The original *Cannon XLR* connector comes apart with three screws. Two hold a strain relief at the back and one holds the insert with the pins. The inserts push out the front. A small screwdriver can help you push it out if it gets tight. The female pin insert can be removed by holding it at the front and pulling it. (See Appendix A, "Notes and Comments," item 27.)

The key advantage of the Cannon XLR over other connector versions is that the female connector has the pins set in neoprene (all others are hard plastic). This allows the pins to "drift" a bit and, when a male connector is inserted, the pins can align themselves, increasing reliability and reducing noise

The Switchcraft A3 is a more streamlined design than the Cannon, but does not offer superior performance. There are two screws at the back that compress an internal ring. This acts as a strain relief. The screw that holds in the pin insert is a reverse-threaded screw. Lose it and you'll never find another like it (unless you cannibalize another connector).

The Neutrik NC3 connector comes from Lichtenstein. Its design is simplicity itself and uses no screws at all. The back of the connector twists off, revealing a three-pronged strain relief. The pin inserts push out to the back and stay held in place by the strain relief when the connector is assembled.

Figure 9-17

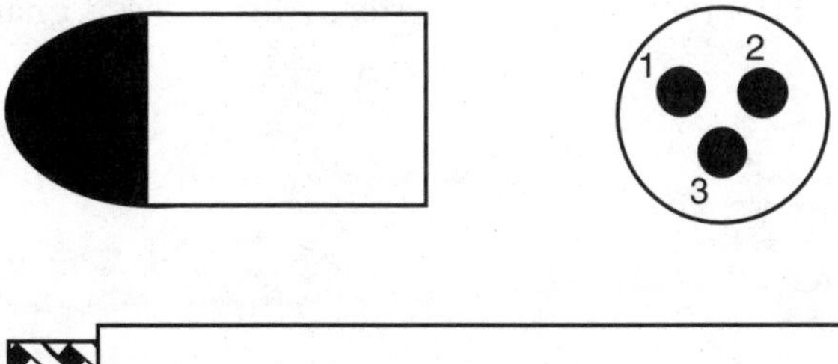

Figure 9-18

All these connectors are compatible. That is, you can plug a male of one into a female of any other. The wiring of them, as far as the pin arrangement, is also identical. Not only are the pin arrangements the same, but every manufacturer has taken the time to number the pins both inside and outside the connectors. All current systems standardize on pin 1 being ground (Fig 9-19).

After all this togetherness, things begin to fall apart because there is no standard for which wire in the balanced line goes to pin 2 and which goes to pin 3 (Appendix A, item 28). This situation is both a problem and a nonproblem, depending on how you approach it.

■ **The Nonproblem** If you are making up microphone cables or snake cables with XLR connectors, and if you do it exactly the same way in every connector, you will be fine. Put the same color wire into pin 2 each time and the same color wire into pin 3. The point here is *consistency.*

■ **The Problem** The problem can start if you're none too careful in making up your cables. If you put the red wire in pin 2 and black wire in pin 3, and at the other end you accidentally reverse red and black, you now have a cable that is *out of phase.*

An out-of-phase cable is one of those things you may never notice until you have a disaster on your hands. Let's say you are recording the president of the United States. Just to be safe, you hook up two microphones pointed at him with two separate cables going into a stereo recorder. Let's further assume you used your out-of-phase cable on one of those microphones.

When you play back the recording, or even broadcast it in stereo, it sounds just fine. But to the listeners who only have a monaural table radio, they would hear virtually nothing. Why? Because the waveform, like the pin wiring on the out-of-phase channel, is reversed with respect to the normal channel. When you combine them (which is what a monaural radio does to the stereo broadcast signals), those two signals

Figure 9-19

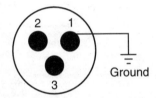

2 1

3

Ground

would cancel each other out. All that you would hear is the very slight difference between the two microphones, a very, very distant president!

If you use that same out-of-phase cable in a recording session, it will have the odd effect of partially canceling the signals from the microphones around it. When they are all mixed together, any sounds that are intended for other mics but happen to bleed into the one out-of-phase will suffer.

Multipair Snake Connectors

Multiple-pair cables, sometimes described as *multicore* cables, can be wired up the same way as other balanced-line cables. If you put XLR connectors on them, just be sure you have pin 1 ground and be consistent about pins 2 and 3.

There are two styles of snake cable, and you should be aware of them and their differences. The traditional style of snake cable takes the individual twisted pairs and twists foil shields around them. Since a foil shield consists of layers of plastic and foil, these shields are usually turned foil-in (toward the twisted pair). The plastic, facing out, can be color-coded so that groups of shields can be identified. Usually the twisted pairs are also color-coded. From the color of the foil and the color of the pair, the order of pairs can be established, and they can be connected in the right order.

This type of snake cable construction allows a compact design. It also allows snake cables to be made in plenum versions. In fact, plenum snake cables are only made in this style.

If you wish to terminate such a snake cable in a multipin connector (often called a "mult"), it is a simple matter of putting the cable through the back shell of the connector, identifying which pair is which, and soldering them correctly. Each foil has a bare drain wire, on which you may want to put some sleeving.

The problem with foil pairs appears when you want to put on separate connectors. The jacket must be removed, exposing all the pairs. Naturally, the foil over each pair begins to unwrap. In order to keep the foil on, you must put on a piece of *shrink tubing*.

Shrink tubing is made with two different kinds of plastics. When heated, the first one melts and allows the other to contract. It is available clear and in colors. Be sure that the "before" size is big enough to fit over what you want covered. In our snake example, you will put a piece of shrink tubing from the point where the jacket was cut to an inch or so from the end of the pair. For that inch, the foil can be removed from the twisted pair and some sleeving put on the drain wire.

You can heat the shrink tubing with a match or lighter or, better yet, with a *heat gun* (Fig. 9-20). These look like a glorified hair dryer, but they can go up to 700°F or even higher. (Don't try to dry your hair with them!) You can only split out the pairs as far as you intend to put on shrink tubing. The tubing is fairly stiff, especially after shrinking, so it is difficult to go more than a foot.

There is a newer generation of snake cable that will make your wiring a lot easier, especially for wiring with XLR-style connectors. Those constructions are called *individually jacketed snake cables*. Each twisted pair is covered by a foil, braid, or serve shield, and each has an individual jacket. All the pairs are then grouped together with an overall jacket.

Individual jackets allow you to split open the outer jacket as far as you want. The pairs are individually jacketed and marked, either by number or color. In some cases, such as Belden's snake cables, they're identified by color, by number, and by that same number written out (e.g., "1-one, 2-two"). Thus you can split out individual pairs as far as you want, even into adjoining racks, without the need for shrink tubing.

Snake Connectors

Snakes are often split into individual XLRs or other connectors. Because of their appearance, they are sometimes called *squids*. Just as often, you can find them attached to a box that contains male and female connectors on the box. There are conduit boxes with entry-clamp arrangements that work well for this. You also might see the box of connectors fitted with one very large multipin connector, which mates to the opposite connector installed on the snake. That way you don't have to drag the box around with the cable, which will coil far more neatly in a road case. The two can be attached to each other when you need them.

One common connector is the Elco or Edac connector. This style of connector, made by two different companies, uses "hermaphroditic" pins. That means the pins are identical in the male and female connector. The pins are most often crimped on the wires, so a special crimp tool is necessary. These connectors have very good reliability, are fairly expensive, and come in a number of sizes and pin counts.

Figure 9-20

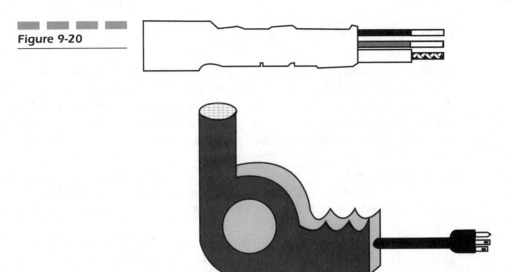

Caution! Because the Elco or Edac connector was not originally made to connect audio snake cables, it does have some size problems, especially with 16-pair snake cable. The shell opening for this size connector is smaller than the outer diameter of most 16-pair snake cables. It definitely cannot fit any 22 AWG snakes, and can fit only one 24 AWG snake (Belden 1916A).

Some installers go to 26 AWG 16-pair snakes (which will fit easily), but 26 AWG is small and may not have the ruggedness required for some installations. Others will use the standard 24 AWG 16-pair snakes, remove the outer jackets, and push only the jacketed pairs through the connector.

There are two reasons why the latter approach is a bad idea. First, removing the jacket dramatically reduces ruggedness. Most cable failure occurs at the strain relief of the connector, the very place the jacket was removed. Second, it looks terrible. Because enough jacket has to be removed so that the wires can be connected and the shell brought down and assembled, as much as half a foot of unjacketed pairs is left exposed. No heat shrink will make this look any better or will add significantly to the lost ruggedness.

Two other connectors often are used for snake cables. The CPC ("circular plastic connector") is made by AMP, Inc., and multipin connectors are made by ITT Cannon (the inventors of the XLR). The AMP CPC has the advantage of being an inexpensive solution for wiring large numbers of conductors.

Speaker Wiring

Speaker wire is usually unshielded and of large gage. There is no standard connector, but there are a few common connectors used.

The first is the *banana plug*. This very simple connector is a reliable connector for large-gage wires. The plugs can be soldered or crimped and the cable can enter the connector from the back or from the side. Often there is a small screw that contacts the bare wire by pressure in the connector and provides a rudimentary strain relief. Banana plugs (and jacks) come in either one-per-wire or in dual configuration (Fig. 9-21). In the dual configuration, there is a tab on one side to show you which lead is the negative (or ground) connection.

Also common is no connector. That is, the wires are separated, stripped, and inserted into terminals that open when pushed and close when released. To prevent stray conductors from shorting to the other terminal, installers sometimes twist each conductor and tin it before inserting it.

Screw terminals are also common. While you can strip and tin the wire and bend it into an appropriate shape, it is more reliable to crimp on a *spade* or *ring connector*. Also called *spade lugs* and *ring lugs*, the best of these have plastic collars over the metal lugs and are crimped on with a special crimp tool (Fig. 9-22).

If the screw is not removable, then bare wire or a spade lug will have to be used. Remember to place the bare wire in the same direction you will be turning the screw (Fig. 9-23).

Headphone Wiring

Headphones are almost always wired with stereo phone plugs that look like, and are wired like, the monaural phone plugs previously men-

Figure 9-21

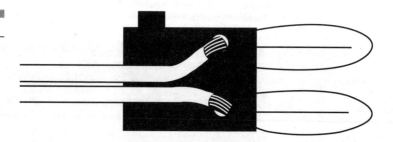

Figure 9-22

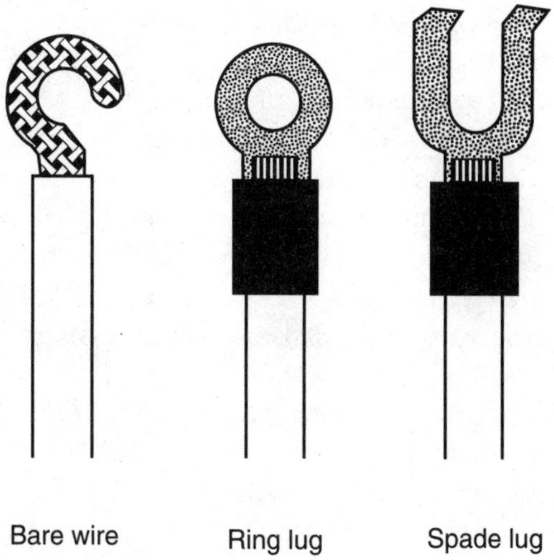

Bare wire   Ring lug   Spade lug

Figure 9-23

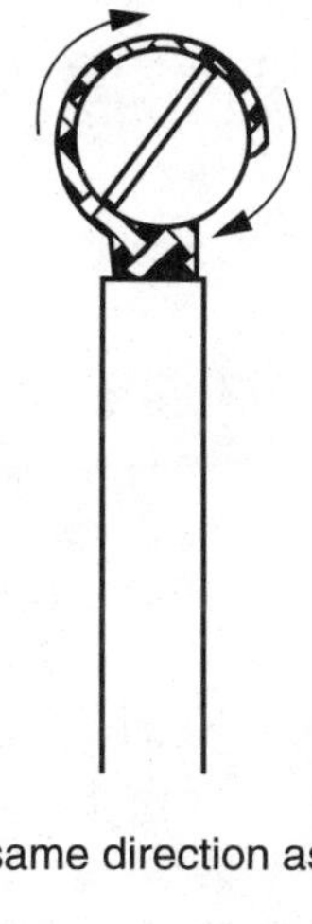

Wire loop in same direction as screw turns

tioned. The exception is there are two tab connections inside, for the left and right positive connections. There is still only one ground connection and it is common to both ears.

Other Audio Connectors

There are many other connectors used for audio. Among them are connectors made by TRW/Cinch, Molex, and AMP. The best way to tell how they are wired is to pull apart a wired one and look at the pin arrangement. If you are keeping an engineering manual (and you should be), it's a good idea to put the wiring scheme for every connector type in it. Copy and distribute those pages to installers or staff who may need to know how to wire these connectors.

Many generic connectors are also making their way into audio, especially subminiature D-shell connectors. These connectors are better known by their pin counts, such as DB-9, DB-15, DB-25, DB-37, and DB-50. These connectors are available in solder versions and "crimp-and-poke" versions.

The solder versions are very hard to work with, especially in the high pin counts, because the spacing is so small between pins. Use very thin solder and an extra-fine tip on your soldering iron.

The crimp-and-poke versions come with separate pins. A special crimp tool is used to crimp the pins onto each wire. They are then inserted ("poked") with a special tool into the connector, where they click into place. This system has proved to be very reliable in the computer world, and most sub-D connectors are installed in this manner.

Video Connectors

There are several types of connectors used for video. For professional use, the most common is the BNC. Other connectors include the F and the RCA (Fig. 9-24).

If premade, off-the-shelf cables are installed using F or RCA connectors, check the cable closely. It can be inferior cable that should not be used for baseband video.

The F style is the standard for CATV wiring. Most premade F connectors use CATV cable, which has a copper-clad steel center conductor and a low-coverage aluminum braid over foil. This construction is appropriate for frequencies over 50 MHz, but not baseband video.

The RCA style is the standard for home hi-fi wiring and is commonly found on consumer or semiprofessional gear. Premade cables with RCA are often intended for audio only, with serve-shielded single conductor cable of unknown impedance. It's also a very poor choice for baseband video, though it's often seen on consumer-grade VCRs.

The BNC connector is the professional standard for video. The best BNCs crimp on the cable. Screw-on or clamp connectors, while easier to put on without special tools, can be less reliable. BNCs are available in 50 and 75Ω versions. The requirements for 75Ω versions for digital video have been discussed in an earlier chapter.

BNC connectors come in two basic styles, *clamp* and *crimp* (Fig. 9-25). The clamp version normally requires that you solder the center pin. There is a clamp at the back of the connector that clamps down on the braid. The crimp-style BNC can also have the center pin soldered, but it is more common to crimp the pin. This requires a special crimp tool. Each size of cable, or change of center conductor, will need a different crimp tool.

Some crimp tools come with removable jaws so that you don't have to buy an entire new tool for a different connector. There are even some crimp tools that can do various sizes of pins and collars, all with the same tool.

Crimping is more common, faster, and can even have improved performance compared to a soldered pin. Skin effect losses at high frequencies are lower with a crimped pin. Solder, with its tin/lead composition, is not as good a conductor. A soldered pin can be more rugged, however, especially with cables that are continuously coiled and uncoiled in the field.

Figure 9-24

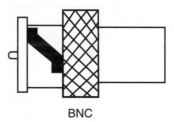

BNC

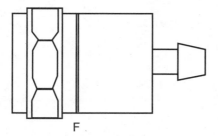

F

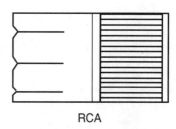

RCA

Figure 9-25

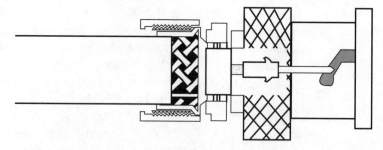

Clamp-style BNC

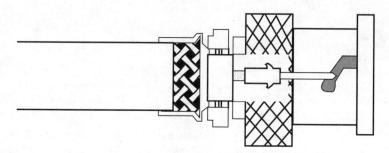

Crimp-style BNC

10

Grounding

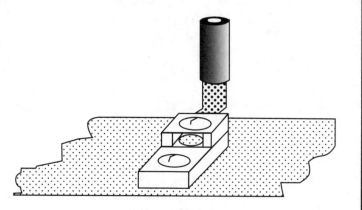

Disclaimer

This is an overview of grounding procedures and practices as they relate to audio, video, data, and other signal wiring. The ground problems considered are those that appear with low-voltage systems, such as noise, radio-frequency interference, and ground loops. The ultimate purpose of this chapter is the preservation of and adherence to the safety aspects of grounding while obtaining the maximum noise reduction.

While electrical safety ground is discussed and analyzed, this book is not intended as a guide to safe electrical grounding procedures and practices. Electrical safety ground design and/or installation should be performed only by experienced, licensed personnel, and accepted and signed-off upon by local building inspectors or similar officials.

Why a Grounding Chapter?

If you talk to an installer about wire and cable, the topic often turns to a discussion of hum and noise. While these can be wire-related problems (with wire-related solutions), they can just as often be ground-related problems. Only the correct design of a grounding system ultimately can solve a ground-caused noise problem.

Virtually all data and broadcast construction projects run into problems of grounding. These problems occur primarily because there is a conflict between issues of safety (i.e., grounding to prevent electrical shock to equipment users) and electronic noise reduction (i.e., using "ground" as an electronic "dump" for noise and interference). These two uses are often not compatible and can sometimes be in direct conflict with one another.

What Is "Ground"?

The definitions of *ground,* physical and electrical, are closely related. The ground, as in the dirt under our feet, has been known for over 150 years to be a good conductor. Early in the history of telecommunications, with the advent of the telegraph, it became known that the earth could substitute for one of the wires needed to make a circuit. This required that the other wire be insulated from ground; it was usually hung on wooden poles, what we would call telegraph poles (or, later, telephone poles). Understanding that the ground could substitute for one of the conductors saved 50 percent of the cost of wiring and a great deal in the labor needed to install it.

Ground and Lightning

Lightning is simply the electrical difference, or *potential*, between a cloud and the ground. A *lightning bolt* is the discharge of that potential difference from the ground to the cloud (although the ionized plasma path of air molecules makes it looks like the lightning is going the other way). Lightning rods give the lightning a very low-resistance path to ground and, therefore, save other structures, vehicles, or people from being in that conductive path. In fact, the idea of a low-resistance path to ground is one of the keys to understanding the uses of ground in electronics.

Ground and the Telephone

When telephones were invented, and later audio equipment (microphones, mixers, amplifiers), it was realized that some of the audio signals were so weak that the slightest electrical noise could interfere with them. To prevent outside interference, a shield of wires was woven around the signal wires. Later, metal foil was also used (see the sections on shielding). For any shield to work, it must be attached to ground. If the connection to ground is a good one, the ground will have a low resistance; the noise will prefer to flow into the ground rather than stay around the signal wires and interfere with them.

Ground and Capacitors

In high-voltage, high-power transmitters, many devices, especially capacitors, store electrical charges that could be lethal to those working on the equipment. Special circuits are installed, which "bleed" away the lethal charges to ground. Sometimes a metal-ended wooden stick is provided, with a wire attached to the metal end (Fig. 10-1).

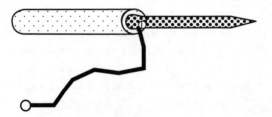

Figure 10-1

The wire connected to the metal rod has a round lug attached to the other end. This must be securely attached to a good ground point. When grounded, an engineer can touch devices and surfaces inside high-voltage equipment. If something is highly charged, the metal tip and wire discharge the voltage to ground, thus ensuring the safety of the engineer. These rods, called *grounding sticks,* are still found around broadcast transmitters and other high-voltage equipment.

Ground and Power Distribution

Meanwhile, with the rise of power distribution to homes and businesses, the lethal nature of even 120 volts (or 220 in Europe) became immediately apparent (see Appendix A, "Notes and Comments," item 19). One way of protecting the user was to enclose plugs, switches, and wires in metal boxes or tubing, called *conduit.* Once the conduit was grounded, then any wiring fault, break, short circuit, or other potentially lethal change would send the electrical power to ground instead of through the user. It is in this context, as safety grounding for electrical power distribution, that ground is most commonly used. Like any other use, however, the ground point chosen (i.e., the connection into the earth) must be a good, low-resistance one. The better the ground, the safer the installation.

A simple safety ground is established by tying a wire to a cold water pipe. Cold water pipes invariably go underground and attach to larger, even deeper pipes. Therefore they make an easy connection scheme for ground. If a portion of metal pipe between the ground wire and the ground itself is broken, missing, or replaced by plastic pipe, however, there is no ground, or an imperfect ground at best. The danger of electrical shock to that building's occupants is seriously increased.

I recall vividly when, as a young boy, I watched workmen in the street installing a new sewer pipe. Down in the hole, the workmen had removed all the dirt around the pipe and had begun to use power saws to cut the old pipe away. They did not cut completely through the pipe, however, leaving one small connection. Then one workman in rubber gloves used a special insulated saw to cut through the last piece. Another workman, also in rubber gloves and holding a meter with two leads, touched one lead to each side of the cut.

It was not until years later that I understood what they were doing. The sewer pipe was probably the main ground connection for all the buildings in that area. If that pipe was an especially good ground, there could be a lot of electrical energy flowing in it at that point. If cutting the pipe had interrupted that flow, the workmen could have been injured or killed by touching both sides of the severed pipe. By using a voltmeter on each side of the pipe, they could tell how energized that section of pipe was. As it turned out, it must have held very little electrical energy, because they proceeded to take off their rubber gloves and cut up the pipe with regular saws.

Regardless of how the safety ground is established, even if it is made by driving copper rods into the ground, there is one absolutely predictable thing that will happen: No matter how good the ground point itself is, by the time you add wire or conduit to that point, and the further that wire or conduit goes from the source, the less protection you will have. This is simply because all wires have resistance; the longer the wire, the more resistance. If there is enough resistance, any electrical fault could find something other than that wire to be a good avenue to ground. If that "other path" is the user, then the safety ground scheme has not succeeded. Most electrical installers try to minimize this by using large-size conduit and, sometimes, by using an ohmmeter to look at the resistance to ground while the building is being constructed.

It should be stressed that the quality of ground at any particular point in a building, compared to some other point in the building, can be dramatically better or worse. The quality of ground at any particular point depends on:

- The size or thickness of the conduit used. (Thicker conduit has less resistance.)

- The type of the joints, elbows, and other connectors used with the conduit. Threaded or welded joints are far superior to the more common setscrew joints used with EMT conduit, simply because they have lower resistance and greater reliability.

- The quality of the workmanship in the conduit installation. Even the best materials can be compromised by shoddy workmanship.

This is a good argument to use a reputable professional electrical contractor in your installation.

■ The quality and reliability of the central ground point to which the building is attached.

Rusted, corroded, or abused grounding rods, dependence on cold-water pipes chosen without regard to their actual ground potential, and high-resistance connections to the building interior can render a ground system next-to-useless.

Ground and the NEC

The National Electrical Code (NEC) devotes an entire chapter to ground. It is in Article 250 of the code book and it addresses almost every kind of ground. The main thrust is personal safety, however, not signal purity. If you're interested in any of the following, get a copy of the National Electrical Code (see Appendix A, "Notes and Comments," item 3).

■ Systems, circuits, and equipment that are required, permitted, or not permitted to be grounded,

■ Location of grounding connections,

■ Types and sizes of grounding and bonding conductors and electrodes, or methods of grounding and bonding, or

■ Conditions under which guards, isolation, or insulation may be substituted for ground.

Other Types of Ground

There are certain installations where running copper pipes or rods is not possible. Planes, boats, and cars are good examples. In those cases, a certain point is designated as the lowest potential point and everything to be grounded is attached to it. It is most often the negative terminal of the battery. Because there are many wires running to that point, often the metal of the vehicle is used and the negative battery terminal is merely attached to the body of the car. Then ground can be established simply by connecting to the body of the car anywhere else.

The steel used in cars (or planes or boats) is not a very good conductor, however. The resistance difference between one point and another can be significant. That will cause a voltage difference between these two points. For lights and other non-signal uses, this is a minor problem. For radios, CD players, microprocessor-controlled subsystems, and other signal devices, this can generate voltage (and noise) between the ground connection of these devices and their positive battery supply connection, reducing their performance or rendering them useless.

Nothing will be as effective as establishing and connecting all your grounds to the actual point chosen, that is, the negative pole of the battery itself. There are a number of wiring devices to allow you to do just that: make multiple connections to ground at your battery. The reduction of ground-voltage and noise pickup can be dramatic. Furthermore, because car battery voltages are low, fewer safety issues are involved.

Safety Ground and Racks

In the data and broadcast world, nearly all equipment is mounted in upright metal cabinets or racks. Doing so makes it easy to establish safety ground. First, the rack itself might be grounded with a large wire running to a central ground point (Fig. 10-2). Second, and most likely,

Figure 10-2

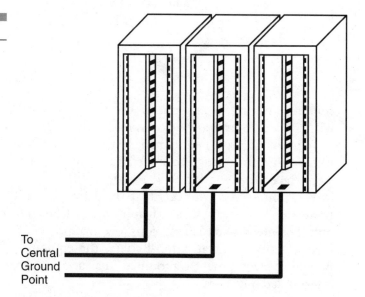

To
Central
Ground
Point

Figure 10-3

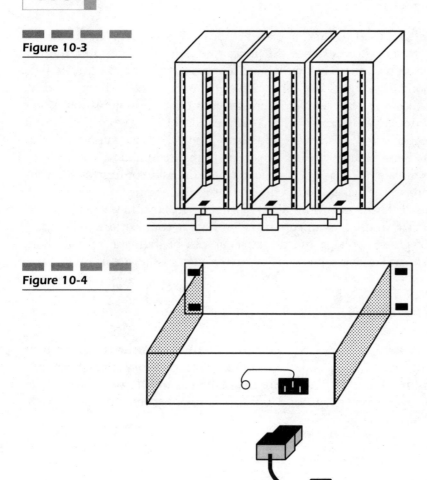

Figure 10-4

conduit running in and out of the rack to deliver ac electrical power will provide ground for the equipment (Fig. 10-3).

Third, virtually all equipment has a three-pin power cord. That third round pin on the plug is connected internally to the metal box or chassis of the equipment (Fig. 10-4).

The power cord connects that third pin to the third pin of the ac power receptacle and, from there to a third green wire inside the conduit. The green wire usually runs to a ground point in a circuit breaker box or similar central point. Because the equipment is grounded, and because the equipment is mounted in a rack with metal screws, the rack is grounded (Fig. 10-5).

And Then There Is Signal Ground

There are other ground circuits going into and out of equipment. There are the shield connections for signals entering and leaving each box in the rack. The shield is grounded to the metal box of that piece of equipment. Shielded cables are often attached to both a transmitting box and a receiving box, so there can be many paths to ground. Figure 10-6 gives you just an idea of all the different ground connections you can find in a group of racks.

These grounds include grounded racks, the conduit attached to the racks, the green wire in the conduit, the power cord that connects the

Figure 10-5

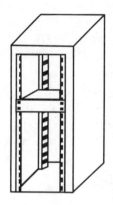

Figure 10-6

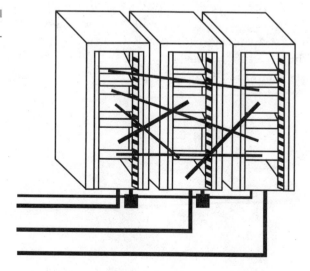

green wire to the equipment, and the input and output cables. While the example shows racks next to one other, it's likely there would be more racks in different rooms, possibly even on different floors, and the signal wiring would be in conduit, trays, or raceways running between these locations.

Ground Potential

Here, then, is the germ of the problem. An audio, video, or data cable could be attached between two pieces of equipment in separate rooms, separate floors, or even separate buildings! If that connection were better than the safety ground at both ends of the cable, or if one safety ground were much better than the other, then significant voltages, noise, and interference could travel down the shields of the signal cables.

Figure 10-7 shows the problem in a schematic form. R_g is the separate ground system used to ground the racks, R_c is the resistance of the conduit from rack to rack (which also could be from room to room), R_w is the green ground wire in the conduit, and R_s is the resistance of the signal ground from one piece of equipment to another. The "arrow" of multiple parallel lines is the schematic symbol for ground.

You can see that the conduit ground is one path, but the signal grounds offer many parallel paths. Each of these paths has resistance (as the schematic shows), and some have more resistance than others.

Part of the confusion in understanding grounding is that there are so many grounds occurring. It will be easier if we can look at one type of ground at a time. For instance, let's look at the ground supplied by the conduit itself. We can simplify our schematic to show the conduit ground as in Fig. 10-8.

Figure 10-7

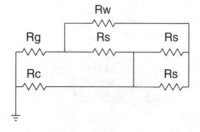

Figure 10-8

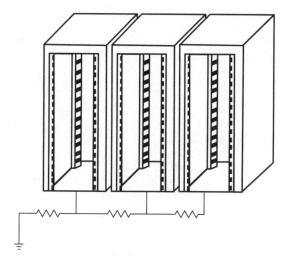

Remember, even the conduit has resistance, and every piece of conduit adds to that resistance. To understand what effect this has, you must understand the idea of a *voltage divider,* which is what this circuit is.

Ground connections can often have voltages flowing down them. These can range from minute RFI or EMI signals, from which the conduit is shielding the wire inside, to lethal voltages from miswiring or accidental faults.

Imagine our racks to be represented by points A, B, and C in Fig 10-9. Let us assume that in rack C there is a voltage on ground for some reason, and that this voltage therefore is on the conduit. This voltage goes through three pieces of conduit (the "resistors") before getting to a central ground point. Now assume there are pieces of equipment in rack A and rack C that are wired together, shown in Fig. 10-10 as resistor X.

It would not take much to make the signal connection between A and C better than the ground connection. You might wonder how a little copper wire could ever be better than a big, fat conduit (even if it is steel). The truth is, it's very easy! If the conduit is rusty or badly installed, a good copper ground wire could easily offer lower resistance.

It also could be that the actual length of the conduit is many feet (maybe even back and forth to a breaker box), while the wire is only a few inches from rack to rack. Finally, there might be more than one wire from A to C. If we were talking about patch panels and routers, there could be a hundred wires from A to C. Even with a short, perfect

Figure 10-9

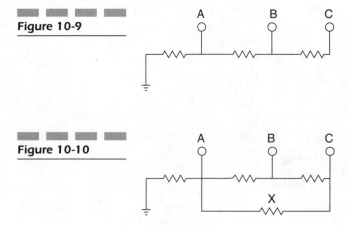

Figure 10-10

conduit installation, this amount of copper is definitely going to be an easier path.

Being an easier path means that the noise, stray voltage, and other electrical garbage will now be traveling down the signal ground, the very thing we were trying to prevent!

Ground is defined as having "zero volts," which indicates that it is the circuit to which everything with any potential other than zero will want to flow. I have seen as much as 42 volts of potential on a shield that should have only microvolts or millivolts of noise. A potential of 42 volts would be tens of thousands times the level of noise normally on that cable. It therefore becomes all but impossible for that noise to be eliminated from the signal path and equipment.

Ground Loops

Once on the signal ground, any noise, stray voltage, or other electronic garbage is fed directly into the equipment that is hoping to be protected from noise. Ironically, it is the shield of the signal cable that enables this to occur, the very thing designed to prevent noise from entering the cable. This connection between safety ground paths and signal ground connections is often called a *ground loop* because there is voltage running in a loop from the power wiring, through the grounding of the cables, and in and out of the equipment.

Even in buildings with excellent ground systems, the run distances can be so great that the natural resistance of the steel in the conduit guarantees a voltage difference from one part of the building to another. Running a shielded cable between these ends virtually assures a ground loop and, therefore, hum or interference of some kind.

You can almost guarantee a ground loop if you are running between two buildings. This is because each building probably has its own source for power from the power company, and each building has its own central safety ground point at the power service entrance. Since this point is not the same point for both buildings, and since the ground in one building is often at lower potential (better) than the other ground point in the other building, there automatically will be a potential difference between the two buildings and, therefore, between any wiring run between the two building.

The effect of this noise varies. In video circuits it shows up as *hum bars,* slow-moving bars that wipe across the screen diagonally. In audio it can be heard as a hum, buzzes, or other constant interface. In data circuits, it can cause bit error rates that can shut down a system in which nothing else is actually wrong.

Quick Fixes

Quick fixes are available for all these problems. They attempt to eliminate the symptoms without addressing the actual problem of poor grounding.

In the video world, one can purchase a *humbucker.* This is a toroidal transformer wound to reject 60 Hz (the frequency of the power line and the most prominent frequency in the noise of a ground loop). Humbuckers do nothing for noise and interference in any other frequency, however, and they are expensive, between $100 and $200 per channel.

For audio wiring, the insertion of an *isolation transformer* can prevent ground loops. Transformers are passive devices that use two coils of wire. While the signal can cross between the two coils magnetically, there is no physical contact. Thus the ground loop is broken. Adding an isolation transformer to each audio line is prohibitively expensive, however, and the quality of even the most expensive isolation transformer does not approach that of a straight piece of audio cable.

Audio equipment often will have a *ground lift switch.* Flipping this switch disconnects the ground connection in this circuit, curing the ground loop problem but also lowering the potential noise-reducing effect of the shield that just got disconnected.

Another quick fix is removing the third (ground) pin from the power cord of one piece of equipment, or inserting a 3-to-2 pin adaptor. If the right unit is chosen, this breaks the ground path because the equipment is no longer grounded. If the equipment is mounted in a rack and the rack or any other piece of equipment in the rack is grounded, however, then that offending piece is still grounded.

Sometimes a lazy technician will remove all the third pins and unground the entire rack to solve a grounding problem. If the specific piece of equipment is identified, however, it can be mounted in the rack with nonconductive washers and plastic screws to prevent any metal-to-metal contact, rendering the piece ungrounded.

While such actions solve the noise problem, removing the third pin disconnects the safety ground and presents a **serious electrical shock hazard**. For the liability reasons alone, this solution should be avoided.

The Fiber Optic Solution

One of the key advantages of fiber optic cable over copper cable is immunity to ground problems. Since fiber is glass, there is no metal contact and, therefore, no ground problem. There are very few machines that are set up to use fiber, however, and those that are usually come at a premium.

Converting to fiber means buying an extra box for each output and each input, a significant added expense. And if your sole purpose is to cure a ground problem, such conversion only cures the problem for those boxes that are converted. All other equipment will still suffer.

It is better to fix the ground problem than try to avoid the problem by going to fiber. The only time fiber may solve an insurmountable ground problem is point-to-point connections or networking between multiple buildings, such as a campus setting, where a common ground point is geographically and logistically impossible.

Medium-Quick Fixes

In the data world they avoid grounding problems altogether by using unshielded twisted pairs (UTP). This has proved to be a wholly acceptable solution. For the analog audio and analog video worlds, however, unshielded cables are not an option. While the data solution is an effective and easy "out," it does not address the actual cause of the problem: poor grounding or a poorly designed ground system.

In general, digital signals are more rugged than analog and can withstand more interference and noise. Thus, converting to digital audio and digital video may help reduce noise problems. The use of UTP for digital audio and digital video is only just beginning to be recognized, however, and there are very few professional, high-quality pieces of equipment that even consider twisted-pair data cable to be an option.

The EIA (Electronics Industry Association) has a study group looking at grounding, but they are between the proverbial rock and a hard place. They cannot recommend that all signal cables be unshielded. In some cases, this is not possible: Hospitals, airports, and other locations with sensitive life-critical systems are very wary of interference getting in, so they often specify shielded wire. Analog audio and video designers do not have an option and must use shielded wire for optimum performance.

Telescopic Grounds

In the audio world there is a solution to ground loops called the "telescopic" ground. A telescopic ground works only with a cable that is a balanced line, i.e., one that has two wires to carry the signal and a separate shield. In a telescopic ground, the shield is connected only at one end. This prevents the completion of the ground loop. It only provides one path for noise and interference to go to ground, instead of two, so the noise reduction is less.

A telescopic ground works best when the end that does have the shield connected to ground is the source end, so any installer must be sure exactly which end he is hooking up. Shield effectiveness gets less and less as you travel further from the grounded end. (That's why the shape of a collapsible telescope, going from large to small, describes the shielding obtained.)

Telescopic grounds cannot be used in unbalanced circuits, such as video coax, because the shield is one of the two conductors necessary to send the signal. That is, the shield is both the noise-reduction portion of the cable and also a signal path. Unground it and—if the signal gets through it at all—you will have the world's noisiest circuit.

The other path, taking the place of the shield that was disconnected, will be established through some other ground path through other equipment! You might as well hang a single wire in midair as an antenna for noise and interference.

The Illegal Solutions

If one could eliminate the safety ground, of course, the problem would go away. But so would the safety of the equipment and, of course, no building inspector in his right mind would sign off on such a practice.

Equally illegal is disconnecting the green wire in the ac receptacle. Also illegal, but less noticeable, would be to leave the conduit as it is and bring the third-pin connections out from the boxes to a separate ground. This means that the safety ground is no longer the conduit but a separate circuit you have established.

Analyzing the Problem

All of these solutions to grounding problems contain one or more of the following flaws:

- They solve the problem for that equipment only.
- They cost a lot of money.
- They trade noise reduction for ground-loop elimination.
- They compromise safety.

What you really need is a system-wide or building-wide solution.

The Real Ground Loop Solution

The key to ground is to have the same potential everywhere and a lower resistance (potential) than any other ground. (That is, if your drain is bigger, offering lower resistance, most or all of the water will flow down it.) The only option, then, is to eliminate the difference in potential between any points wired with cable. If the electrical potential were the same throughout a building, then connecting any of these points to any other points would generate no voltage difference, and therefore no ground loops, hum, or noise.

So the aim in designing a ground system is equal potential at all places. While this may seem impossible, especially when you may have a building with many equipment areas, even split on many floors, it is not impossible and can be realized easily.

The Legal Solution: Star Ground

There are a number of ground schemes intended to eliminate ground loops. The most common is called a *star ground.* In a star ground, a point is chosen as the lowest potential. There are two reasons a star ground works. First, the resistance is lower than any other ground system in the installation. Second, the length of the ground connection between a group of racks and the central ground point is always the same length.

How can you get a low-resistance conductor? Get a big wire! A "big wire" can take many forms:

■ You can separately install a big wire. The bigger the wire, and the lower resistance, the more it will "overpower" every other path to ground and the better it will work. "Big wire" starts at 10 AWG and goes up from there. By the time you're up to 4/0 (0.608 inch outside diameter), it will take some serious money and hefty hardware just to connect it to racks.

■ You can use flat-braid wire, available in a number of sizes. It has very good flexibility and is easy to connect to grounding hardware, but it is expensive. (Braiding is the most expensive step in a wire-making factory.)

■ Perhaps the most popular solution is copper strap. This is bare copper, available in a number of widths (and corresponding resistances). Being wide and thin makes it is easy to bend, turn corners, and attach to hardware. While still expensive, it is probably the cheapest solution when you compare resistance-per-foot versus cost.

One factor that seems to have escaped notice is the requirement for plenum ground wire. Your ground connection will no doubt travel in raised floors or drop ceilings, but I am unaware of any plenum-rated ground wire. If you use bare wire, bare braid, or bare copper strap, however, then there is no plastic on the wire and the metal wire automatically meets plenum requirements. Since the wire is at ground potential, and carries no inherent voltage or signal, it is safe by definition, and need not be jacketed.

You might not want a bare copper strap or wire running everywhere. Remember, anything metal that it touches will be grounded by it. You may wish to avoid air conditioners, motors, and other devices that may introduce noise into your ground strap, thus transmitting it everywhere in your system. This is one advantage of jacketed wire, which will only ground where you strip off the jacket and want it to be grounded.

If you use bare wire or strap, just be careful how you place it. You might even want to get a copy of the building blueprint before you lay the strap and choose the most direct, least-contact path. Don't waste your noise-reduction potential on building items that don't concern your installation.

Your Central Ground Point

The first step in installing a star ground is to run your strap or wire from the building central ground to a central ground point in your own installation. This point is most often a brass plate with multiple holes drilled in it (Fig. 10-11). By having these holes tapped for large brass screws, the arm connections of the star may be established easily.

The strap from the building central ground point should be firmly attached to this block. It can be bolted or soldered to the block. Bolting gives lower resistance initially, but soldering gives a longer-life bond. If you are bolting the strap or wire, use fine sandpaper on the wire or strap and also on the block where the connection will be made. This reduces the corrosion or oxidization to a minimum prior to installation.

Figure 10-11

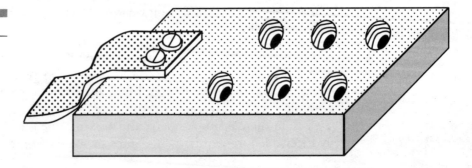

Figure 10-12

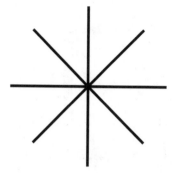

If you are soldering, an iron or soldering gun will be ineffective on something as massive as a strap or brass plate. You will have to use a hand torch, available at any hardware store. Be sure you have sufficient flux with the solder to allow it to flow over a surface this large. For the ultimate in connection, you might consider silver solder, which is the lowest-resistance solder commercially made.

The Arms of the Star

The key to a star ground is that all the ground conductors, or arms of the star, are of equal length (Fig. 10-12). If the ground leads are of equal length, all the ends of the star are, by definition, at the same ground potential. Even though these arms have resistance, if the wire or strap is of an identical size and type in every arm and the distance is identical in every arm, then the resistance of every arm is identical.

This means that the potential between the end of any arm and the end of any other arm is identical. But it must be at each end; you cannot connect halfway down one strap and at the end of another. Only the potential at the ends is identical. Since the potential between any two end points is the same, it doesn't matter how long the arms of the star are as long as they are the same length.

Of course, your installation and your racks are most likely not configured in the pattern of a star. They may be close together or far apart, even on different floors. But as long as you keep the ground connection to each location the same length, the effect of the star ground will be maintained.

Figure 10-13 shows four racks connected in a star ground configuration. The idea that the ends of the star are at equal potential because

Figure 10-13

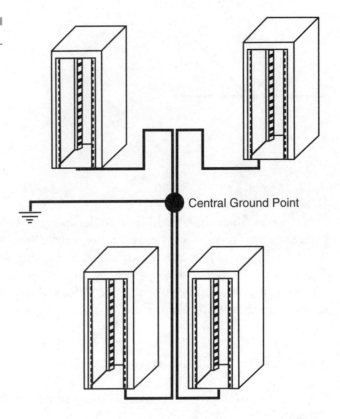

Central Ground Point

they are equal length is critical to the success of this configuration. In fact, length is so critical that, if you have rooms that are both near and far from the central ground point, the farthest point determines the length of the ground conductor to every destination.

This means that a nearby room might have to have a coil of ground wire hidden under the floor or somewhere else, making up for the fact that it is not as far from the central point. Only the end of that wire is connected at the ground point. Care should be taken that the stored loops are either insulated or, if bare wire, do not touch each other or any nearby conducting surfaces. Ground wires may cross the conduit, but should never touch except where each rack is grounded and at the main building ground point. If they touch, the resistance and electrical length will be different than all the other ground wires, the end points will not be at the same potential, and the noise reduction between that rack (or that room) and any other will be compromised.

What's the Standard?

National Electrical Code Article 250-21, which concerns "Objectionable Current over Grounding Connectors" in section D, clearly states that "Currents that introduce noise or data errors in electronic equipment shall not be considered the objectionable currents addressed in this section." In other words, they are concerned only with the safety aspects of grounding, not with noise, hum, ground loops, or data errors.

NEC Section 220-21 b-4 specifically allows an installer to "...take other suitable remedial action satisfactory to the authority having jurisdiction." This means you only have to convince the building inspector or similar authority that your additional ground strap or wire does not compromise the safety aspect of the building ground system. In fact, you should be able to make a good case that the additional grounding actually increases the safety of the installation.

Wire Gage for Grounding

In order that the ends of the star are at very low potential (low resistance), they must be large-gage wire: 10, 8, or 6 AWG is probably the minimum. The AWG number should get smaller (the wire size getting bigger) the farther from each other the points of the star will be. The larger the wire, the lower the resistance and the better the system will work. Generally wire gage is determined less by performance, however, and more by space available, flexibility, and cost. Just realize that anything smaller than 10 AWG will probably give you little improvement over the existing conduit installation. You need to get the majority of signal ground conductivity over your star ground wiring.

Audio Snakes and Grounding

Some of today's snake cables provide not only a drain wire with every pair, but often an overall foil and drain wire with the entire group of pairs. In some installations this addition can be extremely valuable. For instance, if your snake cable is going to a distant location, where running copper ground strap or large-gage ground wire is difficult or impossible, the overall foil-and-drain can be used to establish a "remote" ground potential. Just remember these basic rules:

1. Since this arm of your star is a different length (and resistance) than any other, its purpose is to ground only those pairs within the snake cable itself. If there is a remote junction box (a press box in a stadium, for instance), such a ground can be used quite effectively.

2. If there are signals from other places coming into the remote location where the snake cable terminates, you must supply a star ground arm. The snake shield will not help you.

3. The ground inside a snake cable cannot possibly be as good as a ground strap, or even as good as a piece of conduit. If there are hum and noise problems from that location, a star ground arm must be installed as well.

Copper Strap

It is very common to see the use of copper strap or tape. This is flat copper tape, often two or three inches wide and a quarter-inch thick. Strap can be purchased from hardware stores, plumbing supply stores, electrical whole-salers, or copper-and-brass distributors. Remember that all connections to the star should be the same length. Since the strap is bare, you probably will have to figure out how to store the strap when you have lots of extra. Since it is bare copper, it can be stored in a plenum (drop ceiling or raised floor). Jacketed cable in the same installation would have to have a plenum-rated jacket to be legal.

Determining the Best Ground Point

In each room, care should be taken to choose the absolutely best ground point. The standard is to ground each rack. If racks are bolted together, you may ground the group. It is a good idea to make sure (with an ohm-meter) that there is very low resistance between the ground point and the farthest rack. If you can read anything (even a couple of ohms), it's probably a good idea to ground the rack group at two different points. Just remember, you must run a separate leg off the star, not just extend the strap you have. In only the most critical situations should you con-sider grounding individual pieces of equipment.

Your Ground and Conduit Ground

Remember that even though the conduit is safety ground, you are establishing a second, even better ground. You are not compromising the safety aspects of the conduit. If you use smaller-gage wire or small copper strap, it is possible that the conduit running throughout your installation is of even lower resistance. In that case, the conduit establishes ground, with different ground potentials in each room (and each rack), resulting in noise, hum, and ground loop problems. As noted previously, the key is to overwhelm the resistance of the existing conduit with a dramatically better (lower-resistance) path to the building central ground point.

Critical Ground Points

In especially critical installations that are experiencing ground problems, you can look at the video, listen to the audio, or look at eye patterns or other error-rate data indicators, and move a ground strap around until the image, sound, or data clears up. Wherever you're touching should be the key ground endpoint for that room or group of equipment.

Just to be safe, measure the potential between that strap or ground wire and the rack or piece of equipment before you test it. If there is significant voltage, it means that you must ground that strap or wire to the rack. Then run another, separate piece of strap or wire from that point in the rack to the equipment of concern. That second strap should be long enough to touch every piece of equipment you consider sensitive. The majority of the voltage will be flowing through the grounded rack and strap. Your added path will only show the fine variations in placement of ground for each piece of equipment.

When you are installing the ground strap or wire, if your voltmeter finds a significant voltage difference between safety ground and the signal ground you are touching, all installed equipment should be wired, connected, and plugged in—but turned off. Large voltage transients can permanently damage equipment.

Where and How to Connect the Ground

Most installations bring the ground strap or wire to each group of racks. If the racks are attached together (bolted, for example), they can be considered one unit and have one ground strap. If you've installed more than five racks that have critical circuits in them, even though they're all joined together, you might want to consider a second ground strap.

11

The Installation

A Professional Installation

With the rise of "semiprofessional" audio and video equipment, especially in "project studios" built into homes, often very little thought is given to wiring. Once those installations have grown to even a moderate size, however, the wiring becomes a nightmare. When something goes wrong or you want to modify or rewire a portion of the installation, lack of appropriate planning can waste hours or days. Because of the complexity of even the simplest installation, wiring must be truly professional. Just because the equipment installed is not fully professional doesn't mean it needn't get professional-quality installation.

Professional-quality wiring installation requires only one thing: good planning. This chapter contains many suggestions for a good quality installation.

The System Approach

"Plan the work and work the plan," goes the old adage. So here's the plan:

1. Nothing will help you get to your destination like a map of where you are going. If this is new construction, obtain blueprints of the building or area to be installed. If it is a rebuild or a "project studio," or if you can't get blueprints, make up drawings yourself. Be sure these drawings are made to scale. This will be important when you use it to determine cable lengths.

2. Outline the eventual position of every piece of furniture, every rack, and every conduit (including electrical power conduit).

3. If it's not already shown, add in by hand the conduit, tray, or duct you will be adding to use for wiring. Almost immediately you will see problems:

   a. How are you going to wire that console? How are you going to get the wiring to that edit bay? Raised floor? Conduit coming down from ceiling? Duct in floor? Plenum cable? Nonplenum?

   b. How are you going to wire those monitors or speakers being hung on the wall? Conduit? Or plenum cable from the ceiling?

   c. How can you get from Edit Room C to the machine room when the wall in between is a fire wall?

The Wire List

Figure 11-1 is a chart you can use to guide you through your installation. You probably will want to enlarge it on a photocopier to give you more room to write, and perhaps paste several copies together. Transfer these pages to a master logbook, number the pages, and write in it the following:

1. The project name and the date.

2. FROM: The source device (e.g., "VTR 1," "audio Ch. 1").

3. CONNECTOR: The connector used at the source end (e.g., "XLR").

4. TO: The next termination point of that source, which might or might not be the destination (e.g., "Patch Panel 1 Jack 5-Top").

5. CONNECTOR: The connector at that termination point. If it is punched down instead, write "punch."

6. CABLE: The cable by type (e.g., "analog line") or, even better, by manufacturer and part number (e.g., "Belden 9451/green").

7. LENGTH: Estimate the length of that particular cable. Don't forget that it also might have vertical segments in a rack or wall. When you have the estimated length, add a minimum of 5 extra feet.

Do all the FROM boxes for a particular device in one section. Then do all the TO boxes into that device. That way, when you complete those steps you will find the total for each device. When you are completely finished, you will have two listings for every cable (a FROM listing and a TO listing).

From this chart you will be able to tell a number of things:

1. How many cables you will be installing.

2. What your longest and shortest runs will be.

3. How much cable you will have to buy.

4. What kind and how many of each connector you will have to buy.

5. How many different kinds of cables and connectors you will use.

6. What stripping and crimping tools you will need.

Note that for list item 3 you cannot simply add up the lengths of cables and buy that much. If you need two 600-ft runs (1200 ft total), you cannot buy a 1000-ft and a 500-ft roll; you will need two 1000-ft rolls. Of

Figure 11-1

PAGE:		PROJECT:			DATE:	
Number	From	Connector	To	Connector	Cable	Length

course, other, shorter runs can use up the rest of those spools. Check to see how the cable is packaged ("put up") and arrange your lengths to "fit" with those rolls. It also is a good practice to add 5 ft at each end (10 ft total) for each cable to avoid unanticipated wiring problems.

This list should correspond to any blueprint or other diagrams you have prepared for the project. All of these pages should be included in the master logbook. A three-ring binder is good enough, but be sure to get one that is thick: You're going to put a lot into it. The master logbook should contain all the following:

1. All wire list pages.

2. All drawings of layouts, floor plans, and rack arrangements.

3. All blueprints from contractors or architects.

4. All schedules from designers, contractors, or architects.

5. All contracts with suppliers, distributors, and equipment vendors.

6. All data on equipment you will be purchasing (owners' manuals, assembly instructions, etc.) if not already bound into books.

7. All invoices, bills, quotes, and other paperwork on what you have ordered.

8. All letters, notes, and memos from superiors, coworkers, and staff regarding their desires and requirements in the installation.

When the book is completed, it is a good idea to make a copy and store it in a separate, secure location. You might even want to keep a copy at home. Who knows when you'll be doing another installation for another client or employer and you'll need to know just how you did the last one?

The master logbook will be extremely valuable, especially item 8, when there is any uncertainty, fuzzy memory, or dispute. In 20 years, someone new will probably be looking at these to figure out how everything works.

Labeling

After you have identified every wire, you can begin the process of making labels. No job is as tedious as putting on labels—and no job is so often skipped over. But what good is a wire list if you don't know what wire you're talking about? Spend the time. It will pay off a hundredfold when you are under the gun to find and fix a problem.

There are a number of label manufacturers. Get copies of their catalogs and decide what kind of labels are right for you. Many labels can be printed on a computer. There are even software packages to set up the entire wiring list, from which the labels are generated. They are highly recommended.

The most popular labels are *wraparound* labels. They are partly opaque and partly clear, and they can be written on. They come in a number of different sizes, but generally look like Fig. 11-2. The label wraps around the cable, written area first. As you keep wrapping, the clear area covers the written area, protecting the writing from being rubbed or smudged (Fig. 11-3).

Some people start with a number on the list. The problem with numbering is that you will have to have the list next to you at all times to tell what's what. It is easier to have a list of acronyms or abbreviations to identify sources and destinations. Table 11.1 is a suggested list.

When ordering labels, be sure the opaque area, on which the writing appears, is no greater than the diameter of the cable (Fig. 11-4). If the opaque area is longer than the cable diameter, some of the written text may be obscured by the label as it folds over itself. Ideally, the clear area should cover the opaque area at least one time. The clear area protects the written portion from smudging or otherwise being removed.

Align the label at right angle to the cable so that the clear area covers up the entire written area (Fig. 11-5).

Installers

If you are using a crew of individual installers, you can give each of them a page to work on. (Don't give them your original; make a copy.) You can highlight the exact cables you need done, or you can use different colored markers to determine the order the cables must be pulled and connected (e.g., red first, orange next, yellow next, and so on).

When the installer has completed a page or portion of a page, you might have him or her sign off that page and insert it after the original. That way you will always know who did what. This can be helpful—not so much to lay blame for bad workmanship, but to tell you who to talk to if there are any questions left unanswered, or if expansion in the future is contemplated. Who else but the original installer would know just how something went together, what were the problems and pitfalls that were dealt, with and how best to divide or modify a section for expansion?

Figure 11-2

Figure 11-3

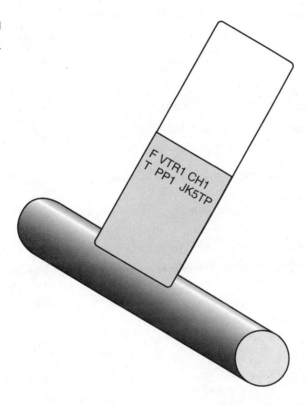

Rack Layout

It is a good idea to make a drawing of your racks, showing the rack divided into single rack units ("RU"). One *rack unit* is $1\frac{3}{4}$ inch of panel height. (Table 11.2 shows a few multiples of this dimension.) Have one picture of every rack. Since there are racks of varying sizes and dimensions, the diagram in Fig 11-6 shows an 84-inch rack, generally considered to be the tallest rack in common use. For equipment that is taller than one RU, naturally you will cover as many units as needed to describe the size of the piece.

TABLE 11.1

Word	Abbreviation
Audio	AUD
Audio Amplifier	AAMP
Audio Cassette Recorder	ACR
Audio Tape Recorder	ATR
Bottom	B
Cable TV	CATV
Computer	CMP
Console	MXR
Digital Audio	DAUD
Digital Audio Workstation	DAW
Digital Video	DVID
Distribution Amp	DA
Edit Suite	EDT
From	F
Graphics	GPH
Ground	GND
In	IN
Jack	JK
Logging Machine	LOG
Machine Control	MCTL
Machine Room	MR
Master Control	MC
Microphone	MIC
Mixer	MXR
Monitor Speaker	AMON
Out	OUT
Patch Panel	PP
Plug	PL
Power Amplifier	PAMP

TABLE 11.1

continued

Word	Abbreviation
Power Supply	PS
Processing	PRC
Public Address	PA
Remote	RMT
Robotic Control	RBT
Router	RTR
RS-232 Control	TTTC
RS-422 Control	FTTC
Satellite Feed	SAT
Stage	STG
Studio	STD
Switcher	SWT
Sync	SC
To	T
Top	TP
Video	VID
Video Cassette Recorder	VCR
Video Monitor	VMON
Video Tape Recorder	VTR

While the dimensions shown are for 19-inch width, there are also 24-inch racks and ones with custom dimensions. Those may require special drawings of their own. You might use a drawing supplied by the rack manufacturer as the template.

It is best to write in pencil, because you will probably want to move equipment around before its final position is determined. If you are having an especially hard time fitting the equipment you will receive into the racks you are planning, try this:

1. Make more copies of the rack layout shown in Fig. 11-6. You might want to enlarge them on the copier to write in more detail, or to draw the location of jacks, signal inputs/outputs, and ac plugs.

Figure 11-4

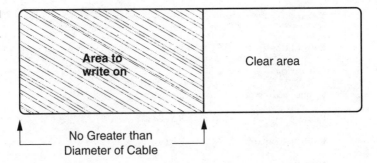

Area to
write on

Clear area

No Greater than
Diameter of Cable

Figure 11-5

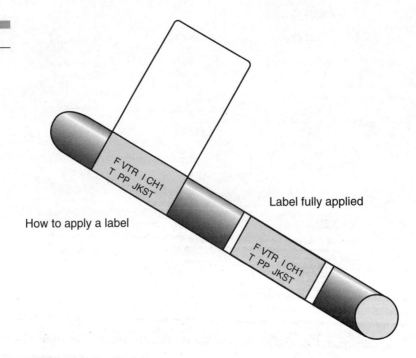

F VTR I CH1
T PP JKST

Label fully applied

How to apply a label

F VTR I CH1
T PP JKST

2. Write on them the manufacturer and model number of each piece of equipment you plan to install. Be sure to use the correct number of rack unit spacings.

3. Using scissors, cut out each "piece of equipment" from the picture.

4. Using other blank rack pictures, play with the pieces until you get them to fit in the order you want.

5. Tape the pieces or transfer the information to the blank racks before you forget.

TABLE 11.2	**Rack Units**	**Inches**
	1	$1\,^3/_4$
	2	$3\,^1/_2$
	3	$5\,^1/_2$
	4	7
	5	$8\,^3/_4$
	6	$10\,^1/_2$

If the wiring inside is looking complicated, use these same drawings but label them as the back of each rack. Indicate the wiring scheme to and from each piece. In this way you should easily identify problems of pieces too close or too far away, or you may find places that will be a wiring nightmare—while other areas are empty and available.

Just remember this Baker's Dozen Rules of Racks:

1. Equipment that will be looked at constantly should be at eye height, between 5 and 6 ft off the floor.

2. Equipment that will be handled and adjusted constantly should be an easy arm's reach, between 3 and 6 ft up.

3. Equipment that pulls out and has adjustments from the top should be no higher than 5 ft up, preferably lower.

4. Equipment that controls or maintains ac power should be isolated from all other equipment.

5. Equipment that tends to radiate or emit RFI should be isolated from equipment prone to pick up RFI.

6. The number of wires in each rack should be approximately equal. Sometimes (such as for patch panels and routers) this is not possible. Just be aware that the more space you provide, the quicker and easier the installation, and the easier maintenance will be.

7. Of the equipment that is left, the heavier equipment should be placed lower in the rack. Power supplies, power amplifiers, and other heavy pieces will help anchor the rack instead of creating a top-heavy hazard.

8. If forced-air cooling from a raised floor blows up through a rack, and it vents through a duct into a drop ceiling, some inspectors

Figure 11-6

might consider the inside of the rack to be a plenum space, requiring plenum cable.

9. Each rack should have plenty of ac receptacles. Count the pieces of equipment from your "unit drawing," and then add at least six receptacles.

10. The receptacle strips should be connected directly to the conduit. This makes the installation safer, neater, and more reliable.

11. There should be one breaker in the breaker panel for each rack. Add up the amperage from each piece of equipment. (See the owner's manuals.) Double the total amps and put in a breaker of that size.

12. Any ac power that is conditioned, backed up by a UPS or generator, or is otherwise different should be run on a different strip (preferably with different-colored receptacles). Put it on the opposite side of the rack from the "normal ac" power strip.

13. Equipment can be kept from "walking away" if it's installed with Phillips-head screws or, better yet, Allen-, Torx-, or hex-head screws. Anyone else will need more than a dime to unscrew that stuff!

Fire Rating Your Installation

On your blueprints, in pencil, determine where the equipment racks will be placed and where the wires will go from there. Does the wire need to go through a wall? Does it go up or down to another floor? Does it travel through a drop ceiling or raised floor? If so, you need to consider fire ratings (National Electrical Code) for the wire you will use. You need to check with your architect, your fire inspector, fire marshal, or permit or planning board to determine what they require. Be aware that the rating is voluntary, so while some areas don't care about ratings, others require even more stringent rules than Code. Most communities today go by the NEC (see Appendix A, "Notes and Comments," item 3).

That means, if you go through walls, most wiring will need to be rated CM. If it is going up (vertically) to another floor, it will need to be rated "riser" or R (as in CL2R, CL3R, CMR, etc.). If it is going under raised floors or over drop ceilings, it needs to be rated "plenum" or P (as in CL2P, CL3P, CMP, etc.). Wiring between pieces in the same rack or between racks in the same room, where it is not hidden from view, usually can be unrated. Check with your fire inspector or permit board.

Putting cable that is unrated, or lesser-rated, into an installation can leave you open to the possibility of liability and fines from any future fire inspector. It is also possible that an unwelcome visit by a fire inspector will have you ripping out wire and replacing it with the correct fire rating. It's better to do it right the first time. When you finally decide what you're going to pull, be sure that every piece meets the required fire rating. If one piece does not meet the correct rating, they all do not.

If you choose cable that is unrated, and you want to go through walls, between floors, or in raised floors or drop ceilings, you have one other alternative: conduit.

Conduit, though accepted by every community for fire rating, is very labor-intensive and expensive. Even with the cost of plenum-rated cable many times the cost of regular cable, it is still much cheaper than conduit.

Conduit has other drawbacks. How do you know you are putting in the right size conduits, or enough conduits, or having them go to the right places? All these can be solved by correct planning. The first step is choosing the right wire.

Choosing Wire

Once you have penciled in the location of your equipment, and have added the wiring runs, and have determined their fire ratings, you must choose the wire to go into those runs. Your basic choices are described in Table 11.3. The most common NEC Fire Ratings are shown. If you need different ratings, consult the manufacturer's catalog. If you need more details on each type outlined below, see the preceding chapters on audio, video, or cable constructions.

Snake Cable

When you pick the appropriate snake cable for your installation, there are a number of considerations. While foil shields are mentioned

TABLE 11.3

Application	Type	Fire Rating	Comments
Microphone	Microphone	Unrated	Flexible, rugged, braid shield (no foil), quad (best), colors
Analog audio	Single pair	CM, CMP	Foil shield, ~30 pF/ft, colors
Analog audio	Multiple-pair snake	Unrated, CM	Foil shield, ~30 pF/ft (see "Snake" below)
Digital audio	Single pair	CM, CMP	Foil shield, 110Ω, < 20 pF/ft
Digital audio	Multiple-pair snake	CM	Foil shield, 110Ω, < 20 pF/ft
Analog video	Coaxial cable	CM, CMP	Single braid, 75Ω, polyethylene dielectric, all-copper center
Precision analog video	Coaxial cable	CM, CMP	Double braid, 75Ω, polyethylene dielectric, all-copper center
Component video	RGB cable	Unrated, CM, CMP	Braid or foil+braid, 75Ω, foam or solid polyethylene dielectric, pretimed
Digital video	Coaxial cable	CM, CMP	Foil+braid, 75Ω, gas-injected foam polyethylene dielectric, all-copper center
RS-232 control	Two twisted pairs	CM, CMP	22 or 24 AWG, two pairs, each foil shielded
RS-422	Multiple twisted pairs	CM, CMP	22 or 24 AWG, 100Ω, equipment determines number of pairs

above, there are also expensive snake versions with braid shields, serve shields, or French braid™ shields. These are all much more expensive than foil shields but offer great flexibility, oxygen-free copper, and other parameters.

If you are installing cable, flexibility sometimes can be a detriment. Flexible cables with soft jackets are often difficult to pull (and especially push) through conduit. What good is the flexibility if your intention is to install it and leave it? As far as oxygen-free copper goes, if you can hear the difference and are willing to pay for it, then buy it.

When determining the number of pairs you will need, consider this. If you need 16 pairs to this machine, 24 pairs to this patch panel, and 32 pairs to this studio from that studio, all these numbers are divisible by 8. Simply buy 8-pair snake. Then use two runs for the 16-channel machine, three for the 24-jack patch panel, four for the 32-channels to the other studio, and so on.

This idea has the added advantage that you buy all your cable in one pair-count, increasing the total volume of snake—and possibly moving you down whole columns in the distributor's pricing. (Instead of 1000 ft of 16-, 24-, and 32-pair, you would order 9000 ft of 8-pair, which should be a lot cheaper per pair.) A further advantage is that you are not left with an odd piece of some odd pair count, too expensive to throw away but of no use to you, or pay a premium for cable cut to length.

Cable Bundles

Now that you've chosen your cable and determined the number of cables from each end to every other end, you must calculate the size of each bundle. You can do this by figuring the cross-sectional area of each cable. Multiply the square of the diameter of each piece by 0.7854 (which is a variation on the πr^2 formula you learned in high-school geometry). Then add the areas together (Fig. 11-7). This will give you a ballpark number for the total area. No bunch of wires will squeeze together perfectly, but it's a rough start.

You can convert back from area to diameter to determine the size of holes in walls, the width between walls for riser installations, and the size of core drilling (if you are putting holes through concrete walls or floors). Simply divide your total area by 0.7854 and take the square root of the result. That will be the diameter of the bundle.

Figure 11-7

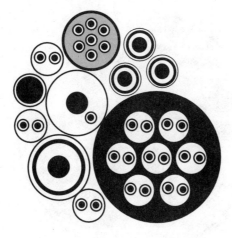

If you intend to put the cable bundle in conduit, the total area you determined can give you a rough idea of the size of conduit to install. Just remember that you cannot fill a conduit 100 percent full. Most installers try to stay below 60 percent full. The National Electric Code standard is 40 percent.

Thus you will understand the disaster that can occur when an architect is told to save money on a design. One of the first things he is likely to do is reduce the size of the conduits. If you have figured for $2\frac{1}{2}$-inch conduit and that is changed to 2-inch, your installation is in trouble. Sixty percent of a $2\frac{1}{2}$-inch conduit is equivalent to 76 percent of a 2-inch conduit, most likely an impossible pull (Fig. 11-8).

Conduit Fill

The NEC standard for pulling cable through conduit is based on a percentage of conduit fill and a number of cables. One cable is allowed to occupy up to 53 percent of a conduit's area. Two cables are allowed 31 percent. Three cables are allowed to occupy 40 percent (see Appendix A, item 25).

You can use Table 11.4 to determine fill or conduit size. To determine permissible fill, look up on the left the conduit size you have. Then go to the column that describes the number of cables you want to put down that conduit. The chart will tell you the percentage of fill and the allowable area. Using the area formulas presented earlier, you can determine the suitability of that conduit (Appendix A, item 20).

Figure 11-8

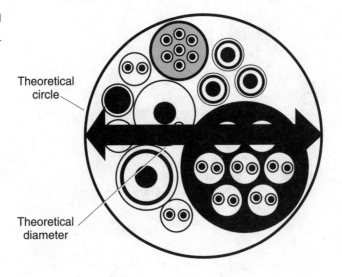

Theoretical circle

Theoretical diameter

TABLE 11.4

Conduit Size (in)	Equivalent Area (in²)	Two Cables 31% Fill Area (in²)	Three + Cables 40% Fill Area (in²)	One Cable 53% Fill Area (in²)
½	0.30	0.09	0.12	0.16
¾	0.53	0.16	0.21	0.28
1	0.86	0.27	0.34	0.46
1¼	1.50	0.47	0.60	0.80
1½	2.04	0.63	0.82	1.09
2	3.36	1.04	1.34	1.78
2½	4.79	1.48	1.92	2.54
3	7.38	2.29	2.95	3.91
3½	9.90	3.07	3.96	5.25
4	12.72	3.94	5.09	6.74
5	20.00	6.20	8.00	10.6

To determine the size of conduit for a group of cables, first determine the total area of those cables (using the method described earlier). Determine the number of cables. Go down the column that applies to that number until you reach an area equal to or greater than the total area you calculated for your cables. Move along that row back to the left until you intersect the "conduit size" column. That number will be the minimum conduit size to use.

Pulling in Conduit

There are a number of tools and techniques that can be used to help speed a pull through conduit. One technique is to count the total degrees of bend. Most common bends are 45° or 90°. Each bend is equivalent to a number of feet of straight pull. Table 11.5 gives the conversion.

The chart ends at the equivalent of two 90° bends, which is a very difficult pull even in a short overall distance. If there are more bends than this, an intermediate pull box should be attached to the conduit; pull the cable to that location and then repull through the rest of the conduit as a separate operation.

Pull boxes also can save you when you wish to modify an installation. They can be a point at which the run can be split into other directions. In some installations, intermediate boxes can be installed where in the future you might expand into an adjoining space. Just be sure the boxes installed have sufficient "knockouts" to expand in the direction you will want to go. Once there's wire running through the box, it cannot be removed, turned, or substituted for by another box.

Just remember that if you're planning to expand in the future, it's a good idea to calculate now the cable you will need for that expansion. You will also need to calculate how much additional conduit you will need,

TABLE 11.5

Total Degrees of All Bends	Divide Fill Percentage by
45	1.2
90	1.3
135	1.4
180	1.5

and determine if you have the space to run it. If you only have enough space for "today", you will never fit "tomorrow."

Metal Underfloor Duct

Many buildings are constructed with metal ducting in the floors. It is somewhat easier to pull cable through than conduit. For one thing, floor ducting often has access plates at regular intervals. This allows you to pull cables in stages, avoiding angled pulls as might be experienced in conduit.

The main problem with metal ducting is that it has to be placed in the floor as the concrete is poured. It takes a lot of planning to be sure the ducting will go under the areas where your equipment will be located. Once it's in, underfloor ducting cannot be added to or changed. You also must have access to pull boxes and end points, even if there is carpet there.

Lubrication and Pull Ropes

There are a number of devices and techniques that can aid in pulling cable through conduit. First is *cable lubricant,* or *pulling compound.* This material is an odorless, yellow-green gel that is inert and nontoxic; it has no effect on cable jackets or conduit. It is very slippery, however, and its sole purpose is to allow the cable to slide more easily. If you have a number of bends or turns, you definitely should buy a few squeeze bottles of this material.

The next tool is a *string blower.* Just a souped-up vacuum cleaner, this device will blow a string down a conduit. You block off the ends of the conduits into which you don't want the string to go. Only the one left open will have air flowing down it, and the string will appear at its end. Once a conduit is full of wire, it is impossible to blow in another string. Having a string in every conduit is a godsend when you need to add that one cable you forgot.

If you don't have a string blower, the next best thing is a *fish tape,* available from any good hardware store. The best fish tapes are self-contained units consisting of flat metal tape on a plastic storage reel. The end can be bent as a hook and can be fished through conduit (or walls or floors). You can either attach a pull string to the tape to start the process, or you can use the fish tape itself to do the pull. Fish tapes are not as strong as you might think, however, and you can easily break the end off as you're pulling. It's better to pull in a string.

It is a good idea to keep a string in every conduit even after the installation is finished. (You'll find out that no installation is ever "finished.") Whenever you use the pull string, attach another piece of string to whatever you're pulling; that way you either pull the original string back in when you're done, or you replace it with a piece of new string. If it's a short pull, tie the two ends of the string together while you're using it. That way you can never lose it. You always have a loop.

The first thing you pull with your pull string is some rope. This rope will pull your cable bundle in. Regular nylon rope is good, or you can buy steel-reinforced nylon. Whatever you buy, check its maximum pulling tension. If you have two or three burly installers, you can easily have 700 lb of pulling tension at the end of that rope!

The next item you will need is a *basket puller.* If you are pulling more than just a few cables, a basket puller is worth the expense. It is a mesh of wire brought to a central point, at which a ring is clamped or welded. The cable bundle is inserted into the wire mesh and a loose wire is woven around the basket, securing the bundle. Duct tape is often added to make the assembly as smooth as possible. The ring at the end makes attaching the pulling rope a breeze—although a refresher course in Boy Scout knots can also work wonders.

How to Pull

Annealed copper begins to elongate or stretch when the tension on it exceeds 15,000 lb/in^2. For multiple cables, or multiconductor single cables, simply determine the gage of each wire inside it and add up the maximum tension for all parts concerned. In a pull of multiple cables, the pulling tension must be equally divided among all the cables. For individual wires, the values shown in Table 11.6 apply.

Just remember that you can easily rip apart an entire bundle of wires by over-eager pulling. Be sure to have one or two people feeding the bundle in one end. If you are using plenum cable, it is especially important that they be able to see into the drop ceiling or raised floor.

Edges of lighting fixtures and mounting brackets, protruding tips of screws, and other sharp objects can rip open cables like a razor blade. If you feel a sudden resistance to your pull, make sure it's not something cutting into your cables.

TABLE 11.6

Wire Gage (AWG)	Maximum Pulling Tension (lb)
24	4
22	7
20	12
18	19
16	30
14	48
12	77

Keep your pulling smooth. Jerking can easily subject the cable to tension beyond the maximum allowed.

Service Loop

Whether you pull your cable with or without conduit, be sure you pull more than you need. You can always cut off the excess. Having too little wastes all your time and all the cable you just pulled. It is good wiring practice not to cut your cables until they have been dressed, i.e., neatened and placed the way they will be permanently tied down. Even then, be careful. If you have a device that slides out for adjustment or routine maintenance, you will want to leave enough cable attached to allow for this extra movement. This is called a *service loop*.

A

Notes and Comments

1 A fringe of scientists believes that the electromagnetic field around the wire *is* the electricity, guided by the conductors, and that there is no other effect occurring in the wire. Electron flow is just an artifice invented to make it all easier to imagine.

2 If only the flow of inventions were this easy. In reality, Alexander Graham Bell invented the receiver, but his transmitter was weak and impractical. Thomas Edison invented the carbon-granule microphone, but his receiver needed to be wound constantly to work, which was not practical either. They fought bitterly in the courts, each trying to get the other essential piece or invent something just as good. They finally compromised, joined forces, and the telephone was born.

3 You can obtain a copy of the NEC book, which lists all the specifications for safe installations in buildings, at any good reference bookstore. You can also call the National Fire Protection Association, which puts out the NEC book, at (800) 344-3555 or (617) 770-3000 to order a copy. (The paperback version is much cheaper!) The sections that pertain to wire, cable, and fiber are summarized in Table A.1.

The top NEC rating is "plenum." In most commercial buildings, cooled or heated air is forced down pipes or vents into appropriate areas. You often can identify these vents as a grille in the center of a drop ceiling. The air then returns unguided by way of the drop ceiling (or sometimes through a raised floor) and is eventually fed through ducts, where it returns to the air conditioning system. It is then cooled or heated again

TABLE A.1

NEC Article	Subject	Comments
725	CL2	Class 2 cables, power-limited but no voltage rating.
	CL3	Class 3 cables, power-limited with a 300-volt rating.
	PLTC	This standalone class is power-limited tray cable, a CL3-type cable that can be used outdoors, is sunlight- and moisture-resistant, and must pass the vertical tray flame test.
760	FPL	Power-limited, fire-protective signaling circuit cable.
770	OFC	Fiber optic cable containing optical *and* metallic conductors.
	OFN	Cable containing only optical fibers.
800	CM	Communications.
820	CATV	Community antenna television and radio distribution systems.

and fed back into the system. Keeping this a closed system saves tremendous amounts of energy. In some buildings the latent heat of the building itself is so great that no heating mechanism is required, only cooling. The air must be kept circulating, however, and, because it must be cooled or heated (thus increasing energy consumption), only a minimum amount of fresh outside air is introduced. The area above a drop ceiling or below raised floor that supplies the return air is called a *plenum*.

If something were burning in a plenum area, any smoke or fumes generated there would automatically be fed throughout the entire building. This is why plenum cable exists. It will not support a fire and, therefore, cannot be the cause of smoke or fumes getting into the return air ducts. For more history on how the National Electrical Code came about, see item 4.

4   Three incidents prompted the National Fire Protection Association (NFPA) to set up the National Electric Code (NEC).

The first incident was a telephone equipment fire on Feb. 27, 1975, in the 11-story AT&T central office switching terminal at 13th St. and 2nd Ave., New York City. The fire started at 12:25 A.M. in the cable vault under the building, and burned for 16 hours. While there were no fatalities, the toxic fumes of burning wire sent 175 firefighters to the hospital. The fire put 170,000 phones out of order in a 300-block area of the city. It required 1.2 billion feet of new service wire and 8.6 million feet of cross-connect wire. It took a 4000-person Bell System task force 22 straight days, working in 12-hr shifts, to restore service, and a total of 562 person-years to rebuild the facility.

In retrospect, it is interesting to consider that 51 percent of the 700 firefighters who fought the blaze never made it to the 20-year mandatory retirement, and that 57 of them (8 percent) got colon cancer.

The second incident was the 1981 MGM Grand Hotel fire in Las Vegas. There was a small fire in the kitchen, which spread to the wiring of the Keno sign. Gamblers refused to leave the tables, or were prevented from picking up their bets by pit bosses, until they were overcome by the toxic fumes of the burning wires. The eventual stampede to the exits prevented many from getting outside. The toxic smoke was drawn into the plenum drop ceiling, where it was pulled out by the air conditioning system and fed to the rest of the hotel. Many jumped from upper floors for no reason; the actual fire was many floors away. They just needed to block their doors or air conditioner vents with wet towels or work their way to the roof. Sixty-seven people died.

The third incident was a rocket attack on a British warship during the 1982 Falklands war between Great Britain and Argentina. Of the crewmen who died, most inhaled toxic fumes caused by burning wires.

5 Oxygen-free copper, or OFC, is copper that has been annealed in a non-oxygen atmosphere to prevent oxygen from rebinding with the copper and allowing oxidization. While the claim is made in high-fidelity circles that you can hear the difference, there has yet to be any laboratory corroboration that OFC makes any measurable difference to the performance of the cable. A purer form of copper will be a slightly better conductor and, at very high frequencies, the skin effect will be slightly lessened. Whether oxygen-free copper makes a significant difference or not has not been proven among most professional audio engineers.

6 The speed of light in a vacuum and in free air is not exactly the same. The air slows it down very slightly, by a factor of 1.0167 to 1. This difference is minimal and usually is ignored. It can be factored into calculations and conversions between dielectric constant, velocity of propagation, and delay. Since we are putting air or nitrogen into the foam, the true formula, and its corollaries, are:

$$D_N = 1.0167 \sqrt{DC}$$

$$D_N = \frac{100}{1.0167V_p}$$

$$DC = \frac{D_N^2}{1.034}$$

$$V_p = \frac{100}{1.0167D_N}$$

where D_N is the delay in nanoseconds, DC is the dielectric constant, and V_p is the velocity of propagation.

It should also be pointed out that the velocity of propagation of glass (as in fiber optic cable) is very poor, much less than the 66 percent of solid dielectrics.

7 There are a number of old broadcast engineers who insist that the origin of 51.5Ω transmission line is not an old wives' tale. They claim that in the 1920s some RCA engineers went to buy 1½-inch and 3-inch copper pipe to make their own line. When they put smaller pipe inside, it came out to 51.5Ω and was called *RMA coax* (Radio Manufacturers of America). When there was enough of a market to justify regular production, the inner conductor was remanufactured to 50Ω.

8 The possible positions of 0 to 1 are configured as an entry bit (0 or 1), the actual bit (0 or 1), and the exit state (0 or 1). This means there are 3 possible states for 0 or 1. Since that is 2 numbers in 3 states, or 2×2×2, there are 8 possible combinations. The eight combinations are listed in Table A.2, along with the sequence in which they would be generated (A through H) by a pseudorandom number generator.

9 Attenuation is a critical number for engineers who want to determine how much power to send to an antenna. A transmitter can be adjusted to add as much power as the cable uses up, thus sending the maximum licensed power to the antenna. In receiving signals, the predicted attenuation may affect how close a receiver is put to an antenna (reducing the cable length and the attenuation it has). This can dictate the use of more expensive larger-size (and lower-loss) cable, or even the added gain of a more efficient receiving antenna, to make up for cable losses.

Usually the engineer uses a chart or graph to estimate the attenuation of a given cable. Especially where the frequency in question is between the points on the chart, or does not appear on a graph, the engineer is reduced to an estimation of the actual number.

In response to this, Dane Ericksen, of the consulting firm Hammett & Edison in San Francisco, has led a one-person crusade for "curve fitting." This technique allows the user to plug numbers into one of three formulas and generate a curve plot of the attenuation for that cable,

TABLE A.2

Entry State	Actual Bit	Exit State	Pseudorandom Order
0	0	0	F
0	0	1	E
0	1	0	H
0	1	1	A
1	0	0	B
1	0	1	G
1	1	0	D
1	1	1	C

often with great accuracy. Each cable design has two variables, which are entered with one of several formulas into an appropriate calculator or computer. The resulting curve plot can be used easily to generate attenuation numbers at any precise frequency.

You can obtain a copy of Mr. Ericksen's paper, which is in the 1989 SBE *Proceedings*, for $20.00 from the Society of Broadcast Engineers, P.O. Box 20450, Indianapolis IN 46220, phone (317) 253-1640.

10 I cannot resist telling what is probably an apocryphal story about grounding. An elderly lady called the phone company to complain of something odd that happened each time the phone rang. Just before her phone rang, her dog, who was chained in the yard, would bark insistently. Then the phone would ring.

After a number of calls to the repair number, the phone company finally sent someone out to investigate this obvious "crackpot." The installer checked the phone and could find nothing wrong. But as he was leaving, he heard the dog barking and then the phone rang! So he went outside to check the dog.

To understand what he found, you have to realize that the phone company sends 50 volts dc to power up each phone and 90 volts ac to make it ring. Often they use ground to establish one side of the ringing circuit (while maintaining a balanced line for the talking part). They establish this ground by inserting a rod in the ground and connecting a wire to it.

In the case of our prescient pooch, there was a telephone pole in the yard and he was chained to the ground post. The post had rusted over the years, however, and was no longer a very good conductor. When the 90 volts came through to ring the phone, it took the path of least resistance—the dog!

Well, 90 volts is a good jolt that would start anybody yelping. Since the dog was not as good a conductor as the metal ground post should have been, the dog got a shock but the phone still did not ring. However, this pooch would get so upset at the number of shocks that it would lose control and urinate on the post. This wet ground suddenly became a good conductor, 90 volts went to ground—and the phone rang! Replacing the ground post (and moving Rover to a less threatening location) solved both problems.

11 Power lost in cable due to changes in resistance can be determined by Ohm's Law and Watt's Law, one of the simplest and most basic set of formulas in electronics. Here are the three formulas in Ohm's Law. They show the relationship between voltage in volts (E), amperage in amperes (I), and resistance in ohms (R):

$$E = I \times R \quad R = \frac{E}{I} \quad I = \frac{E}{R}$$

If any two of those numbers are known, the third can be calculated. Watt's Laws throws in a fourth parameter, power (P), measured in watts:

$$P = I \times E \quad P = I^2R \quad P = \frac{E^2}{R}$$

Just remember that any electrically driven device has an appropriate resistive load on the line. A bad connector or joint adds resistance, which is in addition to the normal load resistance. Therefore the voltage divides between the correct load and the added load of the bad connector or joint, forming what is appropriately called a *voltage divider*. As the bad resistance increases over time (often exacerbated by the heat generated by the resistance at the fault), the division of voltage also increases and the amount of power lost becomes greater and greater until the line melts, a fuse blows, or the piece of equipment driving the line fails. Any seasoned broadcast engineer would proudly show a piece of melted/exploded transmission line caused by this problem. Good connectors and good installation practices keep this occurrence to a minimum.

Many house fires are caused by poor electrical joints overheating. This can either be poor connections, underrated wire, or improper wiring practices. Poor connections can include wire not twisted correctly around screw terminals or fully inserted into push-fit terminals.

Underrated wire can be something like the proverbial extension cord on the Christmas tree. It has a hundred lights on it and they are all plugged into one thin extension cord. When the extension cord melts and short circuits, everyone is surprised.

Improper wiring practices can include mixing metals, such as using aluminum house wire with copper fixtures. Even though they are perfectly installed, the two metals have a different coefficient of expansion when heated. That means that they heat up together but cool down differently, leaving an ever-increasing gap (and an increasing resistance) every time the wire carries electricity. Eventually that gap becomes such a high resistance that the amount of heat it generates melts the fixture and causes a short circuit.

All of these errors result in increased resistance, and increased resistance results in overheated joints. It is not the copper that melts. (Copper melts at over 2000°F.) It is the plastic and other materials that melt first, allowing the conductors to touch and short circuit; the resulting heat causes combustion and a fire.

Appropriate gage size is dependent, therefore, on the melting point of the jacket material. PVC compounds (generally around 60°C) can be made to go as high as 105°C. Other plastics do even better, with Teflon (at 200°C) being the best.

12   In coaxial cable, the impedance is determined by the inductance and capacitance of the cable, or by the ratio of the diameter of the center conductor to the diameter of the shield, with the dielectric constant of the material in between. In twisted pairs, impedance is determined by the gage and distance between the two wires and the dielectric constant of the material in between. Here's one of the formulas:

$$Z_o = \sqrt{\frac{L}{C}}$$

where Z_o is the characteristic impedance of the cable, L is the inductance per unit length, and C is the capacitance per unit length. While the capacitance is often given in the technical data on the cable, the inductance almost never is. Here is another formula, this one based on the diameters of the conductors:

$$Z_o = \frac{138}{\sqrt{E}} \log_{10} \frac{D}{d}$$

Z_o is the impedance of the cable and E is the dielectric constant of the medium between the conductors. For coaxial cables, D is the diameter of the outer conductor (shield) and d is the diameter of the inner conductor. If you play with these formulas long enough, you find that it is very difficult to stray far beyond the typical numbers of coaxial cable. You are limited by the relatively minor differences between dielectric materials, or their foamed cousins. Of course, you can make a very big or very small cable. It's just that making a 20Ω coax, or a 200Ω coax, is very difficult.

13   It is interesting to note that square-wave digital signals are combinations of very high and very low frequencies. The horizontal long sustain of the square is a very low-frequency component. The vertical transitions from 0-to-1 (or 1-to-0) in a theoretically ideal situation are instantaneous and, therefore, of infinitely high frequency. It is at these transitions that the square wave is most susceptible to capacitive storage, since a capacitor (no matter how small) will always block an "infinitely high frequency."

The key in any digital signal is to establish and read the transitions from the 0 to 1 state and the 1 to 0 state. In fact, if all you knew about a digital signal were the transition points, the signal would (theoretically) be fully recoverable.

14 It is amazing how little electrical potential is really needed to send a message by telegraph. In Arthur C. Clarke's wonderful book, *How the World Was One*, he recounts how, once the first two transatlantic cables had been successfully laid, electrical expert Latimer Clark had the far ends connected together (creating a 4000-mile circuit) and was able to send a message using the electricity generated by only a few drops of acid in a lady's silver thimble.

15 The spacing of even a high-coverage braid shield (95 percent) is at least equal to the thickness of one wire. Since a braid is groups of conductors laid across one another, at the transition point the gap is at least one wire thick. If braid is made from 34 AWG conductors (a common size), that is equivalent to 6 mils (152 µm). The actual wavelength of such a braid shield opening is approximately 500 GHz (5×10^{11} Hz). Therefore it is surprising at how low a frequency braid shield begins to fail. In fact, 10 MHz is generally regarded as the point where foil shield effectiveness surpasses braid shield effectiveness. The wavelength at 10 MHz is 30 meters ($98\frac{1}{2}$ ft), so the ratio of that full wavelength to the actual braid gap is approximately 195,000:1. Regardless, a foil-braid solution is essential for any effective broadband shielding.

16 The active-normal digital audio patch panel and Project Patch™ were designed by David Carroll Electronics and are produced by Audio Accessories. These firms can be reached at (510) 528-8054 and (603) 446-3335, respectively.

17 The resistor color code (Table A.3) is increasingly becoming the color code of choice for multipair and multiconductor cable. There are now snake cables in which each pair is individually jacketed and color-coded to the resistor color code. You can tell which pair you are connecting by the jacket color, or by the word or by the number, all printed on the jackets.

Since the resistor color code only goes through ten colors, after that point manufacturers are on their own, using other colors, or colors with stripes or dots, etc. "Color zero" is black and therefore is often omitted from use as a color, because there is no pair zero in paired cable. In some systems, black indicates the number 10 instead of 0.

A black jacket requires an ink color change to mark it (all other colors being markable by black numbers). There are no such limitations to the color black in flat cable, which has no writing on the conductors.

Now it is common for single-pair cable to be available in colors from the resistor color code. This can be useful to color-code installations, either by indicating importance of the signal (red is on-air, green is non-critical, etc.) or coding sources and destinations (all cable from Edit Room 2 is red, all cable from Studio 5 is green, and so on).

18 POTS is "plain old telephone service." One telco employee says they're now playing with PANS, meaning "pretty amazing new stuff." POTS and PANS indeed!

19 It was, in fact, the lethal nature of these voltages that convinced Thomas Edison that ac was the wrong choice for home distribution of power. To prove his point, he would often wire fences with 100 volts or more of alternating current. Pets that wandered into them were, naturally, electrocuted. Edison saw this as a fitting example of the lethal nature of ac, and was not above using it to frighten the public away from his principal competitor, inventor Nikola Tesla.

TABLE A.3

Black	0
Brown	1
Red	2
Orange	3
Yellow	4
Green	5
Blue	6
Violet	7
Gray	8
White	9

Edison wanted to have dc adopted as the power standard; 12 or 24 volts (such as used in motor vehicles) is a very safe voltage, and it is virtually impossible to hurt yourself with such a low voltage. There were two problems, however: resistance and distribution.

If you look at Ohm's Law (item 11), you will see that with a constant voltage (12 volts, say), the size of the wire must increase (therefore reducing resistance) to carry more current. In a toaster, for example, the wire feeding power to it should be big enough so that little is lost in the line, and is instead delivered to the appliance. The power (volts times amps) could be increased on the same wire, however, if you just increased the voltage. It was much cheaper to raise the voltage than to supply a bigger wire.

The other factor was distribution. Former Edison employee George Westinghouse left Edison because he believed ac to be the best way to deliver power. He went to Pittsburgh where, in the 1880s, he was powering electric street cars with ac.

The key to distribution is the transformer, which allows you to trade voltage for current. It consists of two coils of wire wrapped around an iron core. Because the magnetic fields of the two coils cross, the energy introduced in one coil will appear in the other. By changing the ratio of the turns of wire in each coil, you can change what comes out the other side. If it goes from few turns on the entering side (called the *primary coil*) to many turns on the output (called the *secondary coil*), you can make the voltage much higher.

You can't get something for nothing, however. If you step up the voltage, the current will go down, and vice versa. In fact, if you determine the power on both sides (multiply voltage by current), you will find that they are identical (or close to identical, allowing for natural losses in the transformer). You will often hear of power in a transformer expressed in *volt-amperes*, partly because this power is merely transformed, i.e., not being used up or doing work (and partly because of technicalities having to do with the nature of alternating current that are beyond the scope of this discussion).

It was this transformer that Westinghouse used to raise the voltage on his streetcar lines. Thus a small wire could handle a tremendous amount of power. A 1000-volt line at 1 amp was equivalent in power to a 12-volt line of Edison's at 83 amps! A wire that could handle 83 amps was a lot bigger than one that could handle 1 amp. At any place along the system, Westinghouse could put in another transformer that would go from 1000 volts down to, say, 100 volts. Of course, the current would go from

1 amp to 10 amps, so the wire would be bigger from that point on. But it meant that power could be economically delivered down reasonably small wires (at high voltage), and then "stepped down" to a usable voltage at any point along the way.

It is no wonder that power transmission lines are thousands, even hundreds of thousands of volts. Current (and wire size) can be kept low. The only problem is keeping that 10,000-volt line away from people, a problem that persists to this day. So Edison was right, but the almighty dollar (and technology) proved his downfall, at least on this subject.

20   There is a very useful "slide rule"-style chart available from Belden Wire and Cable, called a Conduit Capacity Chart. With it, you can enter the size of your cable and the chart will tell you how many of those cables can fit in a conduit of a chosen size. One side gives Belden part numbers and dimensions; the other side gives generic sizes and dimensions. You can get one free by calling (800) BELDEN-4.

21   The battle between 0.070- and 0.090-inch patch panels is one of those frequent "battles of the giants" that is never resolved. On the 0.070 side was RCA, which had virtually the entire broadcast market sewn up from the 1940s through the 1970s. Because RCA did mostly "turnkey" installations, they decided what got used, and so the 0.070 standard was firmly established. It was later strengthened by patch panel manufacturers such as Trompeter, which also settled on that standard.

On the other side was Western Electric, the largest electronics manufacturer in the world, not to mention sole supplier to the Bell System. Because the Bell System was the main supplier of long lines for video distribution, thousands of 0.090 patch panels were supplied to central offices and other locations where video was being routed and controlled. In network installations, lines leaving and entering the building were often on 0.090 patch panels, confusing matters if that installation was an RCA turnkey project. The 0.090 standard has since been advocated by other patch panel manufacturers, such as ADC.

To be fair, ADC and Trompeter make both 0.090 and 0.070 styles, while others (such as Canare) make only 0.090. Alas, RCA and Western Electric are no more. But their legacy—even their legacy of disagreement—lives on. The 0.090 size seems to have won out, and seems to be used in most new installations. Probably the key reason is that the 0.090 pin is larger and, therefore, more rugged.

22   The manufacturer of digital audio splitters is ETS, (415) 324-4949.

23   The most famous incident regarding a distribution amplifier surely must be the debate between Jimmy Carter and Ronald Reagan. The White House audio crew, and the major networks, had made sure that there were independent backups for everything...except for the distribution amplifier, which fed the resultant audio mix to all participants. As Murphy's Law requires, this was a live, primetime broadcast. When the distribution amplifier failed, there were no feeds to anybody. Eventually, after an agonizing number of minutes, one of the network vans obtained the mixer feed and became the "distribution" source for all other participants. Many actually took the audio feed off that network's local broadcast channel until cables could be wired and strung later in the show. Mercifully, I am unable to tell you who the manufacturer of the distribution amplifier was.

24   While the bandwidth (and therefore potential data rate) of multi-mode fiber is well-known, this is not true of single-mode. The theoretical limit of single-mode was thought to be due to chromatic dispersion. *Chromatic dispersion* is the optical equivalent of group delay in copper cable, where the wavelengths are spread. That is, the fiber is nonlinear or, to be more precise, its index of refraction is nonlinear. This seemed to indicate that there was a maximum bandwidth beyond which the parts of the signal could not be recombined.

One solution is to create *dispersion-limited fiber,* which has been made and even installed in a 70 km run near Tokyo. There is an even more elegant solution, however: Don't change the fiber, change the light!

This solution is called *soliton technology.* Soliton technology divides the optical spectrum up into very tiny channels, as small as 0.2 nm apart in some experiments. When applied to transmission systems, this is called *wave-division multiplexing,* or WDM. Soliton technology also predistorts the pulses so that they stay in shape temporally (time-wise) and spectrally (light-wise) while travelling down the fiber.

The math behind solitons is based on some observations made in the 1830s by a Scottish minister and amateur mathematician, who observed that waves propagated from the sides of barges traveling down the canal he was sitting beside sometimes would hit the shore at an angle and sometimes not. This work was assigned to the "mathematical oddity" department until only a decade ago, when it was realized that his "solitary wave" theory applied to light waves as well.

The current state of the art was shown by—who else?—Lucent Technologies (formerly Bell Labs). In early 1995, Neal S. Bergano, Carl. R.

Davidson, and colleagues demonstrated a WDM system of light pulses that could send eight 5-Gbps channels (that's 40 Gbps!) over 9000 km, equivalent to crossing the Pacific Ocean! They later demonstrated twenty 5-GHz channels (that's 0.1 terahertz!) over 6300 km, equivalent to crossing the United States.

Bruce Nyman, also at Lucent, substituted soliton pulses and also was able to send 40 Gbps over 9000 km. There is belief that the use of soliton technology may open up even greater bandwidths and distances. For instance, a data rate of 10 Gbps has successfully been transmitted over an existing commercially installed submarine cable. It's another Lucent success story. Stay tuned!

25 Quite a few installers are curious why the percentage of fill varies so widely among one, two, and three or more cables. It also is surprising to some that the percentage of fill does not follow that same one-two-three order. The reasons are really quite logical.

First, a single cable is almost perfectly round. Even if "lumpy," it still is more round than even the tightest bundle of loose cables. It also moves as one unit. These two facts account for it having the largest percentage of fill.

Two cables, on the other hand, are the worst shape. They will forever be a figure-eight, which is about as far from round as one can get. Therefore they have the lowest percentage of fill.

Three cables form quite nicely into a triangle, which is a lot closer to a circle than two cables are. More than three gets even closer to a circle. So anything above three is second only to a single cable in allowable fill.

26 One of the best fiber training courses is taught by The Light Brigade, 7639 S. 180th St., Kent WA 98032, phone (206) 251-1240.

27 The story of the XLR and its successors is one that should be required reading for every would-be connector manufacturer. It starts in the late 1940s, when Cannon invented the XLR. Before that, microphone connectors were huge-pinned monsters. (The most popular, the "P-type," was also made by Cannon.) The XLR took the audio industry by storm. Cannon quite literally owned the balanced audio connector market. Competitors, like Switchcraft, didn't have a chance...until 1965. In that year, ITT bought Cannon.

One of the first things they looked at was Cannon's distribution, which was all over the map. Every mom-and-pop electronics store handled Cannon XLRs. ITT saw this as a logistical nightmare and decreed that any distributor would have to carry $50,000 in connector inventory. Almost overnight, the Canon connector disappeared from the audio scene, and competitor Switchcraft, not restricting its distribution, just as rapidly took over. The same occurred in 1987—88 when Neutrik, a Swiss-based connector company, produced a significantly improved version. They, too, have had a dramatic impact on the 3-pin, balanced-line audio connector market.

28   The industry standard for connector wiring seems to be "2=Hot," that is, a positive pressure on the diaphragm of a microphone will cause a positive-going waveform on this wire. Still, there are those who stead-fastly refuse to agree with that assertion. Consider one person in San Francisco, whose auto license plate reads "PIN 3 HOT"!

29   Tables A.4 and A.5 are two charts of data rates and systems. The pro-liferation of different systems, especially those for digital video, has caused much confusion. Just exactly what is "broadcast quality" video? Or what is "VHS quality" video? Table A.4 is a proposed system of levels ("Grade" or "G-levels") that would give a standard set of criteria against which to judge existing broadcast and transmission systems. Table A.5 is an attempt to show all the current data rates, their common uses, the quality to be expected, the medium they use (twisted pair, coax, etc.), and the distances that these systems will run.

TABLE A.4 Video Image Quality Chart

Grade	Bits per Second	Systems at this Data Rate	Pixel Resolution		Frame Rate	Image Quality
G1	<50 kbps	POTS, slow speed	<100 × 200 <(20,000)		<6	Very poor, slow frame rate
G2	50 kbps to <100 kbps	ISDN slow speed, teleconferencing, Switched 56	320 × 200 (64,000) VGA		6-10	Poor rendition of motion. Single frames apparent.
G3	100 kbps to <300 kbps	ISDN multiples	512 × 384 (196,600) MAC		10-12	Obvious frame rate. Severe flicker
G4	300 kbps to <1 Mbps	Partial T-1, teleconferencing	350 × 600 (210,000)		12-15	Pixel refreshing only noticeable with fast screen motion. Moderate flicker.
G5	1 Mbps to <3 Mbps	T-1, MPEG-1, ISDN - end rate	640 × 480 (307,200) SVGA	720 × 484 (348,480) NTSC	15-30	Approaches current home reception quality. Flicker barely apparent
G6	3 Mbps to <10 Mbps	IBM network, Ethernet™ network, DBS single channel	720 × 575 (414,000) PAL	800 × 600 (480,000) VGA	30	Better than current analog home television quality.
G7	10 Mbps to <30 Mbps	Transmitted HDTV	832 × 624 (519,168) MAC		30	Approaches studio quality.

TABLE A.4
continued

			Video Image Quality Chart		
Grade	Bits per Second	Systems at this Data Rate	Pixel Resolution	Frame Rate	Image Quality
G8	30 Mbps to <100 Mbps	DS-3 standard TELCO transport rate	1024 × 768 (786,432) MAC	30	Used extensively for high-quality images
G9	100 Mbps to <300 Mbps	ATM, SMPTE 259M composite/ component broadcast studio quality	800 × 1500 (1,200,000) 1280 × 1024 (1,310,720)	30	NTSC and PAL studio quality
G10	300 Mbps to <1 Gbps	HDTV studio quality, "single-frame-grabber" resolution	900 × 2000 (1,800,000)	30	Approaches detail of film projection
G11	1 Gbps +	Uncompressed HDTV (1.5 Gbps), animation cell rendering, Digital Radar, Digital Micromirror, 6+ Megapixel cameras.	1600 × 1280 (2,048,000) 2560 × 2048 (5,242,880) Radar	30 some 60 frame experimental	Cutting edge, Definition better than film projection

TABLE A.5

Data Rate	Data or LAN Topology	Data Uses	Audio Uses	Video Uses
4.8 kbps	2400 baud	Telephone, FAX, Modem Voice Paging, Data Paging Remote Control, FSK TCP/IP Narrowband PCS	Telephone	
6.4 kbps		Single Digitized Telco Voice-Grade Channel	Telco	
9.6 kbps	4800 baud	FAX, Modem Voice Paging, Data Paging Remote Control TCP/IP Narrowband PCS	approaches AM radio quality Video-Phone	Codec (very slow motion only)
12 kbps				Codec (slowest motion) Teleconferencing Distance Learning
14.4 kbps	7200 baud Category 1 V32 V32-BIS	FAX, Modem Voice Paging, Data Paging Remote Control BIS: Compressed 57K to 14.4 K TCP/IP Narrowband PCS		Codec (very slow motion only) Video-Phone Teleconferencing Distance Learning
19.2 kbps	9600 baud Category 1 RS-232 V32 Turbo	FAX, Modem Voice Paging, Data Paging Remote Control Machine Control RDS, RDF TCP/IP Narrowband PCS	approaches FM radio quality	Codec (very slow motion only) VideoPhone Teleconferencing Distance Learning

For definitions of words, phrases, and acronyms, see the Glossary.

Video Signal Quality	Carrier Transmission Protocol	Medium	Critical Cable Specs	Maximum Distance	Category and/or Variable Systems
	POTS	UTP	None	900 ft. Can be electronically extended up to 6000 ft.	Category 1 V.35 X.25
	Portion of T-1 (DS-1) and higher				
Stop Frame Motion	POTS	UTP	None	900 ft. Can be electronically extended up to 6000 ft.	Category 1 V.35 X.25
Stop Frame	POTS	UTP	None	unspecified	Category 1 V.35 X.25
Stop Frame Motion	POTS	UTP	None	900 ft. Can be electronically extended up to 6000 ft.	Category 1 V.35 X.25
Very poor full motion. T-1 frame rendered in 80 seconds	POTS	UTP	None	900 ft. Can be electronically extended up to 6000 ft.	Category 1 V.35 X.25

TABLE A.5
continued

Data Rate	Data or LAN Topology	Data Uses	Audio Uses	Video Uses
28.8 kbps	V34 V Fast	Fax, Modem TCP/IP		
30 kbps	THRESHOLD—Specific Impedance—Impedance Tolerance			
32 kbps		sample rate	AES/EBU (low)	
44.1 kbps		sample rate	AES/EBU (standard)	
48 kbps		sample rate	AES/EBU (high)	
56 kbps	Switched 56 (S56) DS-0	Remote Audio Wideband PCS		Codec (motion only) Teleconferencing Distance Learning
64 kbps	ISDN (Basic)	Remote audio		Codec (motion only) Teleconferencing Distance Learning
100 k	RS-423	Remote Control Machine Control		
112 kbps	Dual S56			Codec (motion only) Teleconferencing Distance Learning
128 kbps				Codec (motion only) Teleconferencing Distance Learning
250 kbps			Master digital reel-to-reel recording	

For definitions of words, phrases, and acronyms, see the Glossary.

Video Signal Quality	Carrier Transmission Protocol	Medium	Critical Cable Specs	Maximum Distance	Category and/or Variable Systems
	POTS	UTP	None	unspecified	Category 1 V.35 X.25
		STP	110Ω ±20%	unspecified	
		STP	110Ω ±20%	unspecified	
		STP	110Ω ±20%	unspecified	
Poor full motion. T-1 frame rendered in 28 seconds	Switched 56	UTP	100Ω	unspecified	Category 2
Poor full motion. T-1 frame rendered in 21 seconds	ISDN (Slowest speed)	UTP	100Ω	unspecified	Category 2 ISDN
		STP	100Ω	1.2 Km	
Poor full motion. T-1 frame rendered in 14 seconds	Dual S5655 or ISDN	UTP	100Ω	unspecified	Category 2 ISDN
Poor full motion. T-1 frame rendered in 12 seconds	ISDN	UTP	100Ω	unspecified	Category 2 ISDN
		STP	110Ω ±20%	unspecified	

TABLE A.5
continued

Data Rate	Data or LAN Topology	Data Uses	Audio Uses	Video Uses
256 kbps				Codec (motion only) Teleconferencing Distance Learning
336 kbps	Six S56			Codec (motion only) Teleconferencing Distance Learning
375 kbps	Slow QSIF			
384 kbps	¼ T-1 Category 2	Data Transfer		Teleconferencing Distance Learning Compression 370:1
750 kbps	Full QSIF			
768 kbps	½ T-1 Category 2	Data Transfer		Teleconferencing Distance Learning Compression 180:1
1 Mbps	THRESHOLD—Full-Motion Video—Distance Maximums			
1.5 Mbps	Asynchronous SCSI Slow SIF	Network Data Transfer	Audio CD Maximum Data Rate	
1.536 Mbps	Non-channelized T-1	Unsynchronized T-1	Audio STL/TSL	MPEG 1 No interlace "Full motion" compressed ~100:1
1.544 Mbps	T-1 (DS-1) ISDN (full)	Telco Standard		T-1 Extended Super Frame

For definitions of words, phrases, and acronyms, see the Glossary.

Video Signal Quality	Carrier Transmission Protocol	Medium	Critical Cable Specs	Maximum Distance	Category and/or Variable Systems
Poor full motion. T-1 frame rendered in 6 seconds	ISDN or T-1 (partial)	UTP	100Ω	unspecified	Category 2 ISDN
Poor full motion. T-1 frame rendered in 5 seconds	Six S56 ISDN or Partial T-1	UTP	100Ω	unspecified	Category 2 ISDN
Poor full motion					Category 2 ISDN
Mosaic screen refreshing	T-1 (partial)	UTP	100Ω	unspecified	Category 2 ISDN
Mosaic screen refreshing	T-1 (partial)	UTP	100Ω	unspecified	Category 2 ISDN
"Near-VHS"					
"Near-VHS"	T-1 (DS-1) Equivalent to 24 voice channels ISDN (Full)	Specialized Individual Shielded Twisted Pairs	120Ω 100Ω also common	unspecified	Category not applicable but Cat 4, 5 often used ISDN Fastest speed

TABLE A.5
continued

Data Rate	Data or LAN Topology	Data Uses	Audio Uses	Video Uses
2 Mbps 2.048 Mbps 2.052 Mbps	E-1 European Standard			MPEG 1 Fast No Interlace Full Motion compression 72:1 High-quality Teleconferencing
3.072 Mbps			AES/EBU 1-channel transmission standard STP or coax	MPEG 2 Slow Interlace Full Motion compression 48:1 Full SIF Video Duplication Laser Disc Mastering and Duplication (Slow)
4 Mbps	Category 2 IEEE 802.5 IBM System	Network Data Transfer		MPEG 2 Full Motion compression 36:1
4.5 Mbps	Sub 601 (slow)	H.261 slow standard		Teleconferencing Slow Standard
4.8 Mbps				Laser Disc rate
5 Mbps	Synchronous SCSI SCSI II Wide SCSIII "Half D-1"	Network Data Transfer		MPEG 2 Full Motion compressed 29:1 "Half D-1" Laser Disc Mastering and Duplication

For definitions of words, phrases, and acronyms, see the Glossary.

Video Signal Quality	Carrier Transmission Protocol	Medium	Critical Cable Specs	Maximum Distance	Category and/or Variable Systems
		UTP	100Ω	unspecified	Category 2
VHS		STP	110Ω ±20%	100 m.?	
		Coax	75Ω	unspecified	
		UTP	Impedance unspecified	unspecified	Category 3
		STP	150Ω ± ? –6.7dB/M' NEXT 58dB/M'	700 m.	
"Near Broadcast" claimed		Shielded 25-twisted pair parallel or parallel + singles	unspecified	6 m. nominal 25 m. with differential drivers	

TABLE A.5
continued

Data Rate	Data or LAN Topology	Data Uses	Audio Uses	Video Uses
6 Mbps				MPEG 2 Full Motion Satellite DirecTV one channel compressed 24:1 DirecTV HDTV compressed 180:1 CCIR 601 Full (Slow)
6.312 Mbps	DS-2	Telco standard		
8 Mbps				MPEG 2 Fast Full Motion
9 Mbps				1 uncompressed video frame
10 Mbps	RS-422 Fast Synchronous SCSI 2 IEEE 802.3 10 Base 2 Thinnet 10 Base 5 Ethernet Transceiver Drop	Machine Control Network Data Transfer		JPEG 1-Slow Frame transfer 2 frames per second Slow-Speed Networked Video at very high compression Full Fast CCIR 601 Hi-quality H.261 teleconferencing
16 Mbps	Category 3 Category 4 IEEE 802.5 IBM System	Network Data Transfer		Slow-Speed Networked Video at high compression

For definitions of words, phrases, and acronyms, see the Glossary.

Video Signal Quality	Carrier Transmission Protocol	Medium	Critical Cable Specs	Maximum Distance	Category and/or Variable Systems
Better than VHS		STP	110Ω ±20%	unspecified	
		Coax	75Ω	unspecified	
	DS3, equivalent to 96 voice channels	Coax	75Ω	unspecified	
NTSC Single Frame		STP	100Ω	4000 ft.	Category 5
		25- cond Shielded	unspecified	25 m.	
			50 Ω	185 m.	
		Coax			
			50Ω	500 m.	
		Coax			
			78Ω	unspecified	
		STP			
		UTP	100Ω ±15 −40dB/M' NEXT 23dB/M'	800 m.	Category 3
		STP			
			100Ω ±15 −27dB/M' NEXT 38dB/M'	800 m.	Category 4
			150Ω ± ? −13.4dB/M' NEXT 50dB/M'	700 m.	

TABLE A.5
continued

Data Rate	Data or LAN Topology	Data Uses	Audio Uses	Video Uses
17 Mbps				JPEG 1 high-quality frame transfers
20 Mbps	Fast Synchronous SCSI 3 CAM-2	Network Data Transfer		JPEG 1 Fast high-quality frame transfers
25.6 Mbps	Slow ATM	Network Data Transfer		
28.6 Mbps				Proposed 5:1 compression composite video
30 Mbps				Equivalent to one current broadcast TV channel uncompressed
34 Mbps	E-1			
35.75 Mbps	CD-Rom	maximum data storage speed		$^1/_4$ NTSC rate
40 Mbps	THRESHOLD—Broadcast Quality Video—High Definition Television			
44.736 Mbps	T-3 DS-3	Full-Channel Telco Data Link $28 \times T1$		Standard Bell System Video Transmission
50 Mbps	SCSI Short-haul limit			JPEG/MPEG system transfer
51.84 Mbps	OC-1			
54 Mbps				Proposed 5:1 compression component video

For definitions of words, phrases, and acronyms, see the Glossary.

Video Signal Quality	Carrier Transmission Protocol	Medium	Critical Cable Specs	Maximum Distance	Category and/or Variable Systems
Component NTSC Single Frame					
		25 STP & singles			
		Coax	75Ω		
					Category 5
					Category 5
	E-1	Coax	75Ω		
					Category 5
"Broadcast"	T-3 (DS-3) equivalent to 672 voice channels	Coax	75Ω		
	SONET	SM Fiber			

TABLE A.5
continued

Data Rate	Data or LAN Topology	Data Uses	Audio Uses	Video Uses
90 Mbps				Composite High-Definition Television
100 Mbps	Category 5 100 Base T 100 MHz Ethernet Backbone	Network Data Transfer	Hard disc transfer rate	Hard disc transfer rate
143 Mbps				NTSC Composite Uncompressed
155 Mbps	ATM	Network Data Transfer	New specs being written for digital audio on networks	New specs being written for low data rate video on networks
155.52 Mbps	OC-3			
160 Mbps	S-Bus			
177 Mbps				PAL Composite Uncompressed
240 Mbps	EISA			
256 Mbps		Data recorders		Video recording using data format
270 Mbps				CCIR 601 NTSC/PAL Component Uncompressed
274.176 Mbps	DS-4	Telco Standard		
360 Mbps				Proposed HDTV Compressed
466.56 Mbps	OC-9			
622.08 Mbps	OC-12			

For definition of words, phrases, and acronyms, see the Glossary.

Video Signal Quality	Carrier Transmission Protocol	Medium	Critical Cable Specs	Maximum Distance	Category and/or Variable Systems
Broadcast		Coax	75Ω		
		UTP	100Ω ±15 –67dB/M'	100 m.	Category 5
		Coax	NEXT 32dB 50Ω		
Broadcast		Coax	75Ω	540 m.	
	Compression schemes at 31.25 MHz and 77.5 MHz	UTP			Category 5/ Extended Category 5 (DT350)
	SONET	SM Fiber			DT350
Broadcast		Coax	75Ω	480 m.	
					DT350
Broadcast					
Broadcast		Coax	75Ω	405 m.	
	DS-4 equivalent to 4032 voice-grade channels				
1125 line resolution		Coax	75Ω	350 m.	
	SONET	SM Fiber			
	SONET	SM Fiber			

TABLE A.5
continued

Data Rate	Data or LAN Topology	Data Uses	Audio Uses	Video Uses
933.12 Mbps	OC-18			
1 Gbps	Gigabit Ethernet proposed			
1.08 Gbps				HDTV Uncompressed
750 Mbps – 1.20 Gbps +				Motion Picture Frame Rendering. Often transferred at much lower data rates
1.24416 Gbps	OC-24			
1.485 Gbps				uncompressed 16:9 HDTV
1.86624 Gbps	OC-36			
2.48832 Gbps	OC-48			
Unknown				Fractal compression Realtime lossless compression

For definitions of words, phrases, and acronyms, see the Glossary.

Video Signal Quality	Carrier Transmission Protocol	Medium	Critical Cable Specs	Maximum Distance	Category and/or Variable Systems
	SONET	SM Fiber			
	250 Meg × 4-pair	UTP	unspecified	?	DT350
Not in use		Not in use	Not in use		
Beyond Broadcast 2000+ line resolution equivalent		Coax SCSI frame transfer	75Ω	?	
	SONET	SM Fiber			
	SONET	SM Fiber			
	SONET	SM Fiber			
Identical to original	Unlikely before year 2000	?	?	?	?

B

Lampen's Ten Laws of Multimedia

1. A bit is a bit is a bit. Mediums and protocols do not differentiate signal content.

2. Any transmission system can always operate below its maximum potential.

3. No system can produce higher quality than the lowest-quality portion.

4. The quality on any medium is directly proportional to the distance the signal is running.

5. The higher the data rate, the more critical everything is.

6. Anyone who says "It can't be done" is probably wrong. Anyone who says "It can be done" is probably right.

7. Just because it was done that way before doesn't mean it has to be done that way now.

8. The answer is waiting patiently for you to discover it.

9. Nobody really knows anything about the future. We make it up as we go along.

10. Wisdom is not knowing what to do, it's knowing what to do next.

GLOSSARY

AES/EBU Audio Engineering Society (U.S.) and European Broadcast Union (Europe). These two organizations formulated (and continue to formulate) the specifications for digital audio encoding, decoding, and transmission standards.

asynchronous Any protocol in which data is sent at random intervals and the receiving device acknowledges receipt of each signal portion before the next one is sent.

asynchronous SCSI An asynchronous Small Computer System Interface data protocol running up to 1.5 Mbps.

audio Analog signals in the range of audible frequencies, from 20 Hz to 20 kHz. In digital audio, any portion of this spectrum which is sampled and digitized.

balun Literally meaning "balanced to unbalanced," a balun is a device for converting from balanced cable (UTP, STP) to unbalanced cable (coax) or vice versa.

baud In data transmission, 2 bits or state changes in the transmission medium, roughly equivalent to one cycle of analog information.

bit One state transition in a digital signal (such as off to on to off or 0 vdc to +5 vdc to 0 vdc) that transmits one piece of information.

byte A sequence of 8 bits of data.

CAM 2 In SCSI networking, the "Common Access Method 2," which allows users of different operating systems to access SCSI devices in a consistent way, with few or no changes in the protocol.

CCIR 601 The standard digital video protocol formulated by the European Broadcast Union.

Cheapnet Nickname for an Ethernet network wiring protocol based on standard RG-58 coaxial cable.

codec Meaning literally "code and decode," a codec is a system designed to send video images, both still and motion, using medium- to low-speed data pathways (thus reducing cost of transmission). The codec scheme looks at the picture in motion (such as someone talking) and changes only that part of the frame that is moving (such as the speaker's mouth). With very slow-speed data, the in-motion parts may be redrawn only every second or more, not even close to full-motion but good enough for teleconferencing, video-phone, and similar low-quality transmission.

compression   Any scheme that processeses a digital signal to reduce the number of bits, while retaining the original information and making it decodable. Generally, the greater the compression, the more errors will occur during the decoding process and the lower the quality of the resulting product. Compression is usually expressed as a whole number. For instance, "3-to-1 compression" means one bit is left where originally there were three. Some compression schemes perform better than others at certain tasks (such as digital audio or video).

DS-1   An encoding standard used by common carriers (i.e., telephone companies), with a data rate of 1.544 Mbps. The transmission for this standard is called T-1.

DS-3   An encoding standard used by common carriers (i.e., telephone companies). The data rate is equivalent to 28 DS-1 channels, or 44.736 Mbps. The transmission standard for this is called T-3.

E-1   An encoding standard used by common carriers (i.e., telephone companies). The data rate is 20 DS-1 channels, or approximately 34 Mbps. E-1 is used for cross-connecting and patching, and therefore has no corresponding transmission standard.

Ethernet™ A networking protocol invented by Xerox, originally coax-based but now also carried on UTP. As implemented for coax, Ethernet is a bus (linear) topology; typically a terminal has a **T** connector, the leg of which is connected to the terminal. The other two ports (the crosspiece of the **T**) are part of the bus, and have cables coming in and going out. The twisted-pair implementation of Ethernet, called "10Base-T," is a star topology in which each terminal is wired with a "home run" to a hub or concentrator.

As in any baseband network, only one terminal at a time can transmit on the network. In the Ethernet protocol, terminals initially attempt to send their data packets without first checking for network traffic. If more than one terminal starts transmitting simultaneously, "collision" detection circuitry will stop both; each terminal will wait a random amount of time before trying again.

Fast SCSI   A "high-speed" version of SCSI, capable of 10 Mbps data transmission.

Fast-Wide SCSI   The "top-speed" version of SCSI, with 16-bit (instead of 8-bit) data words. Fast-Wide SCSI can reach data rates up to 40 Mbps.

FAX   The transmission of "facsimile" or copies of written or drawn documents. Documents are scanned in a line-by-line format and reproduced in that fashion by the receiver. Time for transmission is determined by the data rate, usually expressed in kbps or Mbps.

55 Octet   Multilevel transmission scheme proposed by the Advanced Video Group of Pacific Telesis to carry 6 Mbps full-motion "VHS-quality" video over standard POTS lines. "Octet" is a telco term equivalent to one byte.

fractal compression   Proposed scheme of "lossless compression" using fractal encoding; the resulting image would be identical to the original source material. The encoding and decoding is so mathematically intensive that fractally compressed real-time video is not likely to be seen before the year 2000.

frame   A single, still video image; 30 frames per second is considered full-motion video. Frames can be of low or high quality, depending on how much data is used to render or draw them.

FSK   Frequency-shift keying, a very simple way of encoding relatively slow data by shifting between two predetermined audio tones. This method can pass down virtually any transmission medium, including voice-quality phone lines. FSK is simple and slow but electronically robust.

FTP   File transfer protocol, one of the suite of Internet protocols that governs the movement of files from host to host over a network.

full motion   A video display rate of 30 frames per second.

IBM   International Business Machines, a manufacturer of computers and data systems. IBM often specifies special or nonstandard cable types for many of its systems.

IEEE   Institute of Electrical and Electronic Engineers, an engineering organization that also has committees within the organization to set standards for various data systems and formats.

impedance   A cable parameter that determines the cable's ability to pass energy efficiently from one end to the other. Generally, the impedance of the source and receiving device should be the same as that of the wire or cable. Impedance becomes more critical as bandwidth, frequency, or data rate rises.

ISDN   Integrated System Digital Network, a system used by common carriers and local phone companies that allows digital "calls" to be placed to any location set up for ISDN, much like dialing an analog ("POTS") phone. ISDN bandwidth, which can be adjusted as needed for the application being run, ranges from 64 kbps to T-1 speed (1.544 Mbps).

JPEG Joint Photographic Experts Group, an imaging industry organization that has established a standard video frame transfer protocol.

k A prefix to a unit of measure that multiplies the quantity by 1000.

K A quantity of computer memory or disk storage equal to 1024 (2^{10}) bytes, not to be confused with the lowercase k, which multiplies by 1000 (10^3). Both forms are found in computer usage, depending on context and the nature of the quantity being described.

kbps Kilobits per second, a unit of data transmission speed equal to 1000 (10^3) bits per second.

LAN Local area network, a combination of two or more computers, terminals, printers, storage devices, or other devices on a common wiring scheme, allowing data transfer from or to any of the network devices.

m Meter, an SI unit of length equal to approximately 39.37 inches.

M A prefix to a unit of measure that multiplies the quantity by 1,000,000.

Mbps Megabits per second, a unit of data transmission speed equal to 1,000,000 (10^6) bits per second.

medium The material over which a signal is traveling. It can be copper, such as multiconductor, twisted-pair, or coaxial; glass, such as fiber optic; or wireless, such as radio or microwave.

meg, mega Prefix meaning "one million."

mosaic refreshing A technique, used in some *codecs*, that attempts to hide the fact that the device cannot keep up with a high data rate. In a series of video frames in which major portions of the frames are in motion, the fast-motion portions are reduced to mosaics or squares of very low data quality, which fade in and out. As the image slows and the data catches up, the squares fade into smaller and smaller squares until the maximum data rate (and picture detail) is reached.

MPEG Motion Picture Experts Group, a working group within the SMPTE who set, among other things, specifications for compression schemes for video transmission.

MPEG 1   A compression scheme set up by MPEG, intended for still photographs but used for many other purposes.

MPEG 2   A compression scheme set up by MPEG to cover full-motion video in many formats, from teleconferencing to HDTV.

NEXT   Near-end crosstalk, the amount of signal that travels, via induction, from an energized pair into an unenergized pair in multipair bundles or combinations. This is measured at the source ("near end") where the signal is strongest, so the crosstalk is given as a worst-case condition.

NTSC   National Television Systems Committee.

PCMCIA   Personal Computer Memory Card International Association, a computer industry group that defined standards for portable, nonvolatile memory cards (also called "smart cards") typically used in laptop or notebook computers.

POTS   "Plain Old Telephone Service," carried on standard telephone paired cable, with no specifications, no tolerances, and no performance requirements.

RAID   Redundant Array of Inexpensive Disks, used to describe disk storage systems that distribute data across multiple, off-the-shelf disk drives to increase digital storage capacity and reliability.

remote audio   Sending audio on a temporary basis from one location to another for broadcast. The technique usually is used to send high-quality voice audio back to the studio for live remote broadcasts from conventions, churches, fairgrounds, sporting events, etc.

RG   "Radio Guide," a military specification system that set out basic parameters for building certain types of cables. Each specification was numbered, e.g., RG-6, RG-174, etc. Cables specified as "Type" (e.g., RG-58 Type) are a variation on the original military specification.

SCSI   Small Computer System Interface, a protocol for connecting computers and peripherals that uses cables with multiple pairs and single conductors. Originally used strictly with disks for digital archiving, SCSI now is used also for data transfer between computers and a variety of peripherals. Though simpler and faster than Ethernet, SCSI lacks true networking capabilities and is limited to very short bus lengths.

SMPTE   Society of Motion Picture and Television Engineers, an organization of engineers within the film and video communities.

SMPTE contains working groups that set standards such as JPEG, MPEG, and 259M.

SONET Synchronous Optical Network, a telephone company high-data-rate transmission protocol based on single-mode fiber optic cable.

S-DIF Sony Digital Interface Format, a single-channel consumer digital audio standard based on coaxial cable and RCA connectors.

SP-DIF Sony/Phillips Digital Interface Format, a one- or two-channel consumer digital audio standard based on coaxial cable and RCA connectors.

STL Studio-to-transmitter link, a series of devices to send audio, video, and/or remote-control data from a studio to a transmitter site, where it will be broadcast or used for transmitter control. *See TSL.*

STP Shielded twisted pairs.

S-VHS "Super video home system." An analog system to increase the quality of VHS recording, editing, and playback. In cable, it uses pairs of coaxes to separately carry the luminance (black-and-white) and chrominance (color) picture information.

Switched 56 A transmission system offering 56kbps data transfer. The "switched" refers to the ability of the phone company to route the data to many locations, almost like dialing a phone.

T-1 The transmission scheme of DS-1, with 1.544 Mbps data rate, 23 voice channels ("B channels") and 1 signal circuit ("D channel").

T-3 The transmission scheme of DS-3, at a spped of 44.736 Mbps. Thirty T-1 circuits equal a T-3.

TCP/IP Transport Control Protocol/Internet Protocol, a suite of protocols for various types of transactions on data networks and the Internet. Implemented by the Department of Defense in the 1970s for ARPAnet (precursor of the Internet), TCP/IP in recent years has become widely used by the general public in local and wide area networks. *See X.25.*

Thinnet A small, coax-based networking cable based on RG-58.

TSL Transmitter-to-studio link, a series of devices to send audio, video, and/or remote-control data from a transmitter site back to a control point (studio) for control, monitoring, and security purposes. *See STL.*

UTP Unshielded twisted-pair cable, usually referring to Category 3, 4, or 5 premise cables.

V.32bis   A slow-speed data transmission protocol.

VHS™   Video Home System, a low-quality consumer video delivery system used for recording and playback of video signals. VHS is a trademark of Panasonic, Inc.

video   Images produced by electronically scanning and projecting line-by-line. Each image or frame can be a still, or can be in multiples to create slow- or full-motion. Full motion video is 30 frames per second.

Wide SCSI   SCSI that uses 16-bit words instead of the 8-bit words of standard SCSI, allowing increased data transfer with the same clock frequency.

X.25   The public wide-area, packet-switched network standard. Competes with the TCP/IP protocol.

INDEX

Illustrations are in **boldface**.

About the Author

Stephen H. Lampen is a Technology Development Manager at Belden Wire and Cable. He had an extensive career in broadcasting, is an SBE Certified Radio Broadcast Engineer, and holds an FCC General Lifetime License. On the network/data side, he is also a BICSI Registered Communication Distribution Designer (RCDD).